Mathematical Lectures from Peking University

Mathematical Lectures from Peking University includes monographs, lecture notes, and proceedings based on research pursued and events held at Beijing International Center for Mathematical Research (BICMR). BICMR is a mathematical research institute sponsored by the national government of China. The center was created in 2005 by national government decree. Its goal is to build a world-class mathematical center for research and training young talents; to foster a new generation of leading world-class mathematicians in China; to support the application of mathematics in sciences and practical fields; to promote research and improve the level of mathematics education at Peking University, as well as all over China.

More information about this series at http://www.springer.com/series/11574

Zhibin Liang · Chandrakant Aribam
Editors

The Computational and Theoretical Aspects of Elliptic Curves

Springer

Editors
Zhibin Liang
Capital Normal University
Beijing, China

Chandrakant Aribam
Department of Mathematical Sciences
Indian Institute of Science Education
and Research Mohali
Sahibzada Ajit Singh Nagar, Punjab, India

ISSN 2197-4209 ISSN 2197-4217 (electronic)
Mathematical Lectures from Peking University
ISBN 978-981-13-6663-5 ISBN 978-981-13-6664-2 (eBook)
https://doi.org/10.1007/978-981-13-6664-2

Library of Congress Control Number: 2019932699

This Springer imprint is published by the registered company Springer Nature Singapore Pte Ltd.
The registered company address is: 152 Beach Road, #21-01/04 Gateway East, Singapore 189721, Singapore

Foreword

The Birch and Swinnerton-Dyer (BSD) conjecture is a central conjecture in the arithmetic of elliptic curves. It was postulated in the 1960s with numerical evidence to support the plausibility of this conjecture. In its history of about sixty years, it has seen some progress and has also led to fruitful, new developments in arithmetic geometry. Though the conjecture in its entirety is still far from being solved, the conjecture has lured many researchers to attempt a solution. The problem itself is a classic example of being amenable to theoretical advances as well as numerical computations that support the theoretical developments. Further, advances in computational algorithms have played an important role in the last five decades in explicit numerical calculations. The conjecture has also been generalised to the arithmetic of modular forms and has opened up connections with other areas of arithmetic geometry such as Iwasawa theory.

This volume is an attempt to present a collection of results related to the BSD conjecture, based on the first two India–China conferences on this theme. It presents an overview of the conjecture and some cases where the conjecture is proved. The broad theme of the two conferences was 'Theoretical and Computational Aspects of the Birch and Swinnerton-Dyer Conjecture'. The first one was held in Beijing International Centre for Mathematical Research (BICMR), Beijing, in December 2014, and the second was held at the International Centre for Theoretical Sciences (ICTS), Bangalore, India, in December 2016. It is hoped that this volume will benefit younger researchers who wish to work in this area, as it provides a broad overview on the subject. The articles have an extensive list of references, and the diligent researcher can pursue these works to get an idea of the current state of the art for this conjecture.

The organisers of the two workshops and the editors of this volume are grateful to the two institutions, BICMR and ICTS for providing support and hosting the workshops. The administrative staff in both these institutions went beyond their call of duty to ensure extensive participation as well as providing assistance on all matters related to logistics and practical arrangements.

Sujatha Ramdorai
December 2018 Member, Organizing Committee

Contents

Introduction to the Conjectures of Birch and Swinnerton-Dyer

Sudhanshu Shekhar and R. Sujatha

The aim of these lectures is to introduce the Birch and Swinnerton-Dyer conjectures in its entirety. One part of these conjectures predicts the equality of two different 'ranks' associated to an elliptic curve defined over a number field. These are the so called algebraic and analytic ranks. The other part of the conjectures is an exact formula expressing the leading coefficient of a certain power series associated to the elliptic curve in terms of various important and mysterious arithmetic invariants. The approach we shall take is to define and provide a brief introduction to these arithmetic invariants, thereby providing a compact introduction to this conjecture. We omit the details, referring the interested reader to Silverman's book [1]. Other excellent references to the theme of this article are [2, 3].

We thank BICMR Beijing and ICTS Bangalore for the excellent hospitality and the referee, whose report helped improve the exposition of the article. The first author would like to thank Department of Science and Technology (DST) for the support of INSPIRE grant. The second author gratefully acknowledges the support of NSERC Discovery Grant 402071.

Weierstrass equation of elliptic curves. An elliptic curve over a field K is a projective non-singular curve of genus one defined over K with a specified base point (see [1, Chap. I, Sect. 2] and [1, Chap. II, Sect. 5]). Recall that any such curve E has a (Weierstrass) equation of the form

$$Y^2Z + a_1XYZ + a_3YZ^2 = X^3 + a_2X^2Z + a_4XZ^2 + a_6Z^3 \tag{1}$$

S. Shekhar (✉)
IIT Kanpur, Kanpur, India
e-mail: sshekhars2012@gmail.com

R. Sujatha
University of British Columbia, Vancouver, BC, Canada
e-mail: sujatha@math.ubc.ca

Z. Liang and C. Aribam (eds.), *The Computational and Theoretical Aspects of Elliptic Curves*, Mathematical Lectures from Peking University,
https://doi.org/10.1007/978-981-13-6664-2_1

in $\mathbb{P}^2$, the projective space of dimension two, with $a_1 \ldots a_6 \in K$ for $1 \leq i \leq 6$. Here, $O = [0, 1, 0]$ is the base point (called "the point at infinity"). By dehomogenising (i.e. taking $x = X/Z$ and $y = Y/Z$) the above equation can be expressed as

$$y^2 + a_1xy + a_3y = x^3 + a_2x^2 + a_4x + a_6. \tag{2}$$

Thus $E \subset \mathbb{P}^2$ consists of the points $P = (x, y)$ satisfying the above Weierstrass equation along with the base point O. If $char(K) \neq 2$, then we can simplify the equation by the change of coordinate

$$y \mapsto 1/2(y - a_1x - a_3)$$

which gives an equation of the form

$$E : y^2 = f(x) \tag{3}$$

where

$$f(x) = x^3 + b_2x^2 + 2b_4x + b_6,$$

and

$$b_2 = a_1^2 + 4a_4,\ b_4 = 2a_4 + a_1a_3,\ b_6 = a_3^2 + 4a_6.$$

Associated to the above equation we define the following quantities

$$\begin{aligned}
b_8 &= a_1^2a_6 + 4a_2a_6 - a_2a_3^2 - a_4^2,\\
c_4 &= b_2^2 - 2b_4,\\
c_6 &= -b_2^3 + 36b_2b_4 - 216b_6,\\
\Delta &= b_2^2b_8 - 8b_4^3 + 9b_2b_4b_6,\\
j &= c_4^3/\Delta,\\
\omega &= dx/(2y + a_1x + a_3) = dy/(3x^2 + 2a_2x + a_4 - a_1y).
\end{aligned}$$

An easy verification shows that,

$$4b_8 = b_2b_6 - b_4^2 \text{ and } 1728\Delta = c_4^3 - c_6^2.$$

If $char(K)$ is different from 2 and 3, then using the change of coordinates

$$(x, y) \mapsto (x - 3b_2/36, y/108)$$

Equation (2) can further be expressed as

$$y^2 = x^3 - 27c_4x - 54c_6.$$

Substituting $A = 27c_4$ and $B = -54c_6$ we get that the equation

$$y^2 = x^3 + Ax + B$$

which is usually called the *short Weierstrass form.* Put $\Delta = -16(4A^3 + 27B^2)$. The quantity $\Delta \in K$ is an important invariant associated to the curve E, called the *discriminant* of E over K. Further, the non-singularity of E implies that $\Delta \neq 0$ and the cubic $f(x) = x^3 + Ax + B$ has distinct roots. The quantity j is called the $j - invariant$ of the elliptic curve, and ω is the *invariant differential* associated to the Weierstrass equation.

For a field extension L/K we define the set

$$E(L) := \{(x, y) \in L^2 |\, y^2 + a_1xy + a_3y = x^3 + a_2x^2 + a_4x + a_6\} \cup \{O\}.$$

It is well known that for any field extension L/K there exists an abelian group structure on $E(L)$ such that

1. O is the identity element with respect to this group structure.
2. If $L_1 \overset{\phi_{L_1,L_2}}{\longrightarrow} L_2$ is a homomorphism of field extensions of K, then there exists a corresponding group homomorphism

   $$E(L_1) \overset{\phi^\star_{L_1,L_2}}{\longrightarrow} E(L_2)$$

   defined as

   $$\phi^\star_{L_1,L_2}(x, y) = (\phi_{L_1,L_2}(x), \phi_{L_1,L_2}(y))$$

   satisfying $\phi^\star_{L_1,L_2}\phi^\star_{K,L_2} = \phi^\star_{K,L_2}$.

In particular, if L/K is a Galois extension then $E(L)$ is a $\mathrm{Gal}(L/K)$-module. The above group structure can be geometrically described by the well known chord-tangent method (see [4, Chap. I]).

Elliptic Curves Over the Complex Numbers.

Definition 1 A *lattice* in the field of complex numbers $\mathbb{C}$ is a discrete group of the form $\Lambda = \mathbb{Z}\omega_1 + \mathbb{Z}\omega_2$, where ω_1 and ω_2 are linearly independent over the real numbers $\mathbb{R}$. Two lattices Λ and Λ' are said to be *equivalent* if there exists $\lambda \in \mathbb{C} - \{0\}$ with $\lambda\Lambda = \Lambda'$. A complex torus T is a quotient $\mathbb{C}/\Lambda$ of the complex plane $\mathbb{C}$ by a lattice with projection denoted by $p : \mathbb{C} \longrightarrow T = \mathbb{C}/\Lambda$.

Remark 2 If $\lambda \in \mathbb{C} - \{0\}$ such that $\lambda\Lambda \subset \Lambda'$ for lattices Λ and Λ', then it induces a homomorphism $\lambda : \mathbb{C}/\Lambda \longrightarrow \mathbb{C}/\Lambda'$. Such a map is called a *homothety* induced by λ. A homothety $\lambda : \mathbb{C}/\Lambda \longrightarrow \mathbb{C}/\Lambda'$ is an isomorphism if $\lambda\Lambda = \Lambda'$. In fact, it can be shown that every complex analytic isomorphism between two tori is associated to a homothety.

Definition 3 *An elliptic function f with respect to a lattice* Λ is a meromorphic function on $\mathbb{C}$ such that $f(z+w) = f(z)$ for all $z \in \mathbb{C}$ and $w \in \Lambda$.

The *Weierstrass $\wp$-function* associated to a lattice Λ is given by the infinite sum

$$\wp(z;\Lambda) = \wp(z) := \frac{1}{z^2} + \sum_{\omega \in \Lambda - \{0\}} \left[\frac{1}{(z-\omega)^2} - \frac{1}{\omega^2} \right]. \tag{4}$$

The *Eisenstein series* of weight $2k$ associated to a lattice Λ is the series

$$G_{2k} = G_{2k}(\Lambda) = \sum_{\omega \in \Lambda - \{0\}} \omega^{-2k}. \tag{5}$$

Theorem 4 ([1, Theorem 3.1 and Theorem 3.5, Chap. VI.3])

(a) The Eisenstein series $G_{2k}(\Lambda)$ is absolutely convergent for all $k > 1$.
(b) The series defining the Weierstrass $\wp$-function converges absolutely and uniformly on every compact subset of $\mathbb{C} \setminus \Lambda$. This series defines a meromorphic function on $\mathbb{C}$ having a double pole with residue 0 *at each lattice point and no other poles.*
(c) The Weierstrass $\wp$-function is an even elliptic function.

The Laurent series for $\wp(z)$ around $z = 0$ is given by

$$\wp(z) = \frac{1}{z^2} + \sum_{k=1}^{\infty} (2k+1) G_{2k+2} z^{2k}.$$

Furthermore, we have

$$\begin{aligned}
\wp(z) &= \frac{1}{z^2} + 3G_2 z^2 + 5G_3 z^4 + \cdots \\
\wp'(z) &= \frac{-2}{z^3} + 6G_2 z + 20 G_3 z^3 + \cdots \\
\wp'(z)^2 &= \frac{4}{z^4} - \frac{24G_2}{z^2} - 80G_3 + \cdots \\
4\wp(z)^3 &= 4\wp(z)\left(\frac{1}{z^4} + 6G_2 + 10G_3 z^2 + \cdots\right) \\
60G_2\wp(z) &= \frac{60G_2}{z^2} + 180 G_2^2 z^2 + \cdots .
\end{aligned}$$

Comparing the first few terms of the above expressions, one sees that for all $z \in \mathbb{C} \setminus \Lambda$, the Weierstrass $\wp$-function and its derivative satisfy the relation

$$\wp'(z)^2 = 4\wp(z)^3 - 60G_2\wp(z) - 140G_3$$

Put $g_2 = g_2(\Lambda) = 60G_4(\Lambda)$ and $g_3 = g_3(\Lambda) = 14G_6(\Lambda)$.

Proposition 5 ([1, Proposition 3.6]) *Let $g_2 = g_2(\Lambda)$ and $g_3 = g_3(\Lambda)$ be the quantities associated to a lattice $\Lambda \subset \mathbb{C}$ as above.*
(a) The polynomial

$$f(x) = 4x^2 - g_2x - g_3$$

has distinct roots, so its discriminant

$$\Delta(\Lambda) = g_2^3 - 27g_3^3$$

is non-zero.
(b) Let $E/\mathbb{C}$ be the curve $E : y^2 = x^3 - g_2x - g_3$, which from (a) is an elliptic curve. Then the map

$$\phi : \mathbb{C}/\Lambda \longrightarrow E(\mathbb{C}) \subset \mathbb{P}^2(\mathbb{C})$$

$$z \mapsto [\wp(z), \wp'(z), 1],$$

is a complex analytic isomorphism of complex Lie groups, i.e. it is an isomorphism of Riemann surfaces which is a group homomorphism.

For an elliptic curve E over $\mathbb{C}$, let $E[m]$ denote the subgroup of m-torsion points of E defined over $\mathbb{C}$.

Corollary 6 *For every integer $m \geq 1$, there is an isomorphism of abelian groups $E[m] \cong \mathbb{Z}/m\mathbb{Z} \times \mathbb{Z}/m\mathbb{Z}$.*

Corollary 7 *Let E_1 and E_2 be two elliptic curves corresponding to lattices Λ_1 and Λ_2 as in the above proposition. Then E_1 and E_2 are isomorphic over $\mathbb{C}$ if and only if Λ_1 and Λ_2 are homothetic.*

An important theorem in the theory of elliptic curves over $\mathbb{C}$ is the $Uniformization\ Theorem$, which asserts that every elliptic curve E defined over $\mathbb{C}$ corresponds to a lattice Λ_E as in the above proposition, i.e. there exists a lattice Λ_E uniquely determined up to homothety such that

$$\phi : \mathbb{C}/\Lambda \longrightarrow E(\mathbb{C})$$

$$\phi(z) = [\wp(z, \Lambda), \wp'(z, \Lambda), 1]$$

is an isomorphism of complex Lie groups and E has a Weierstrass equation given by

$$y^2 = 4x^3 - g_2(\Lambda)x - g_3(\Lambda).$$

The lattice Λ_E for an elliptic curve E given by a Weierstrass equation as in (2) is, in fact, the set of periods

$$\int_\gamma \omega_E$$

where γ runs over all closed paths in $E(\mathbb{C})$ and ω_E is the associated invariant differential $\frac{dx}{2y+a_1x+a_3}$. Equivalently, Λ_E is the image under the homomorphism

$$H_1(E(\mathbb{C}), \mathbb{Z}) \longrightarrow \mathbb{C}$$

$$\gamma \mapsto \int_\gamma \omega_E$$

where $H_1(E(\mathbb{C}), \mathbb{Z})$ denotes the homology group of $E(\mathbb{C})$ with coefficients in $\mathbb{Z}$. The lattice Λ_E is also called the *period lattice* associated to the curve E. A generating set for Λ_E can be obtained by integrating ω_E over a basis $\{\gamma_1, \gamma_2\}$ of $H_1(E(\mathbb{C}), \mathbb{Z})$ (see [1, Chap. VI, Proposition 5.2] for more details).

Elliptic curves over local fields. Let K be a perfect local field, complete with respect to a discrete valuation v and R be the ring of integers of K. Let $\mathfrak{m}$ be the maximal ideal of R, π be a uniformizer of K, i.e, a generator of $\mathfrak{m}$ and denote the residue field of R at $\mathfrak{m}$ by k. Further, we normalize the valuation v such that $v(\pi) = 1$.

Let E/K be an elliptic curve, and let

$$E : y^2 + a_1xy + a_3y = x^3 + a_2x^2 + a_4x + a_6 \tag{6}$$

be a Weierstrass equation for E/K. By substituting $(x, y) \mapsto (\pi^{-2t}x, \pi^{-3t}y)$ for a sufficiently large integer t, we may assume that the $v(a_i) \geq 0$ for the coefficients a_i as in (6). In particular, this implies that the valuation $v(\Delta)$ of the discriminant Δ associated to the above equation is ≥ 0. Further, since v is discrete, we can choose a Weierstrass equation defined over R which minimizes the value $v(\Delta)$. Such a Weierstrass equation of E is called a *minimal (Weierstrass) equation* for E. The minimal value of $v(\Delta)$ is called the *minimal discriminant* of E at v.

Proposition 8 ([1, Proposition 1.3, Chap. VII]) *(a) Every elliptic curve E/K has a minimal Weirstrass equation unique up to a change of coordinates*

$$x = u^2x' + r, \quad y = u^3y' + u^2sx' + t$$

with $u \in R^\times$ and $r, s, t \in R$.
(b) The invariant differential,

$$\omega = \frac{dx}{2y + a_1x + a_3}$$

associated to a minimal equation is unique up to multiplication by an element of $R^\times$.

Given a minimal Weierstrass equation of the form (6), we can reduce its coefficients modulo π to obtain a curve over k given by

$$\tilde{E} : y^2 + \tilde{a}_1 xy + \tilde{a}_3 y = x^3 + \tilde{a}_2 x^2 + \tilde{a}_4 x + \tilde{a}_6 \tag{7}$$

The curve $\tilde{E}$ is called the *reduction of E modulo π*. The restriction of the reduction map from

$$\mathbb{P}^2(K) \longrightarrow \mathbb{P}^2(k)$$

induces a map

$$E(K) \longrightarrow \tilde{E}(k)$$

which is again called the *reduction map*. If the curve $\tilde{E}$ is non singular, then E is said to have *good reduction* over K, otherwise E is said to have *bad reduction* over K. Let $\tilde{E}_{ns}(k)$ denote the set of non singular points of $\tilde{E}(k)$. In particular, if E has good reduction over K then $\tilde{E}(k) = \tilde{E}_{ns}(k)$. If $\tilde{E}$ has bad reduction over K then the following situations occur:

(i) If $\tilde{E}$ is a cuspidal cubic, then $\tilde{E}_{ns} \cong \mathbb{G}_a$, and E is said to have *additive reduction over K*.
(ii) If $\tilde{E}$ is a nodal cubic, then $\tilde{E}_{ns} \cong \mathbb{G}_m$, and E is said to have *multiplicative reduction*. Two further sub-cases occur in this situation which we mention now. If the tangent directions at the node of $\tilde{E}$ are defined over k then E is said to have *split multiplicative reduction over K*, otherwise it has *non-split multiplicative reduction over K*.

Put

$$E_0(K) = \{P \in E(K) | \tilde{P} \in \tilde{E}_{ns}(k)\}$$

$$E_1(K) = \{P \in E(K) | \tilde{P} = O.\}$$

Proposition 9 *The sets $E_0(K)$, $E_1(K)$ and $\tilde{E}_{ns}(k)$ have a group structure such that $E_0(K)$ and $E_1(K)$ are subgroups of $E(K)$. Further, we have the exact sequence*

$$0 \longrightarrow E_1(K) \longrightarrow E_0(K) \longrightarrow \tilde{E}_{ns}(k) \longrightarrow 0.$$

Here, the right hand map is the reduction map modulo π.

Definition 10 The *Tamagawa number* of E over K is defined as the index $c_K(E) := [E(K) : E_0(K)]$.

Theorem 11 (Kodaira, Néron) *Let E/K be an elliptic curve. If E has split multiplicative reduction over K, then $E(K)/E_0(K)$ is a cyclic group of order $v(\Delta) = -v(j)$. In all other cases, the group $E(K)/E_0(K)$ is finite and has order at most* 4.

If E has split multiplicative reduction over K, then by a theorem of Tate

$$E(K) \cong K^\times / q_E^{\mathbb{Z}}$$

where q_E is called the *Tate period* of E over K, and is related to the j-invariant $j(E)$ of E via the equation

$$j(E) = j(q_E) = q_E^{-1} + 744 + 196884 q_E + 21493760 q_E^2 + \cdots .$$

Here, j denotes the modular j-function (see [5] for more details, see also [6, Chap. V, Theorem 3.1(b)]). In this case the Tamagawa number $c_K(E) = ord_v(q_E)$ (see [1, Corollary 15.2.1]). If E has good reduction over K then $c_K(E) = 1$.

Elliptic curves over number fields. If K is number field and E/K is an elliptic curve then we have the following celebrated "Mordell-Weil theorem" (see [1, Chap. VIII]).

Theorem 12 *If K is a number field and E/K is an elliptic curve defined over K, then $E(K)$ is finitely generated as an abelian group.*

In particular
$$E(K) \cong \mathbb{Z}^{r_K(E)} \oplus E(K)_{tor}.$$

Here, $r_K(E)$ is called the *rank* of E/K and $E(K)_{tor}$ is the (finite) torsion subgroup of $E(K)$.

For a non-archimedean prime v of K let k_v denote the residue field at v. We say that E has good (resp. bad) reduction at v if E has good (resp. bad) reduction over the completion of K at v. For a non-archimedean prime v, we define the integer

$$a_v(E) := q_v + 1 - \#\tilde{E}(k_v)$$

where q_v is the number of elements in the finite field k_v.

Definition 13 The local L-factor of the Hasse-Weil L-function of E at v is the polynomial defined as

$$L_v(E/K, T) = \begin{cases} 1 - a_v(E)T + q_v T^2 & \text{if } E \text{ has good reduction at } v \\ 1 - T & \text{if } E \text{ has split multiplicative reduction at } v \\ 1 + T & \text{if } E \text{ has non-split multiplicative reduction at } v \\ 1 & \text{if } E \text{ has additive reduction at } v. \end{cases}$$

The Hasse-Weil L-function of E over K is defined by the Euler product

$$L(E/K, s) = \prod_v L_v(E/K, q_v^{-s})^{-1} \quad \text{for } Re(s) >> 0$$

where the product varies over all non-archimedean primes of K.

By results of Hasse and Deligne, if v is a prime of K where E has good reduction and
$$1 - a_v(E)T + q_v T^2 = (1 - \alpha T)(1 - \beta T)$$

then $|\alpha| = |\beta| = \sqrt{q_v}$, where $|\ |$ denotes the complex norm. Thus, $|a_v(E)| \leq 2\sqrt{q_v}$. This, in particular implies that the the above Euler product converges in the right half plane $Re(s) \geq 3/2$.

Conjecture 1 *For an elliptic curve E over a number field K, the Hasse-Weil L-function of E has an analytic continuation to the entire complex plane $\mathbb{C}$.*

As a consequence of the Modularity theorem proved by Wiles et al. and the theory of base-change for automorphic representations of $GL(2)$ (cf. [7–9]), we have the following deep

Theorem 14 *Let E be an elliptic curve defined over $\mathbb{Q}$ and K be a solvable Galois extension of $\mathbb{Q}$. Then the Hasse-Weil L-function $L(E/K, s)$ has an analytic continuation to the entire complex plane.*

The order of vanishing of $L(E/K, s)$ at $s = 1$ is called the *analytic rank* of E over K.

Periods of elliptic curves. Now, consider an elliptic curve E defined over a number field K. If the class number of K is one, then it is possible to find a Weierstrass equation which is simultaneously minimal at all non-archimedean primes of K. Such an equation is called a *global Weierstrass minimal equation* of E. Suppose that

$$E : y^2 + a_1xy + a_3y = x^3 + a_2x_2 + a_4x + a_6$$

is a global Weierstrass minimal equation of E over K. Then, the *real period* of E is defined as

$$\Omega_{E/K} := \int_{E(\mathbb{R})} \frac{dx}{2y + a_1x + a_3} \in \mathbb{R} \tag{8}$$

Note that $\Omega_{E/K} \in \Lambda_E$ where Λ_E is the period lattice associated to E. If we assume Conjecture 1, then there is a uniformly convergent power series expansion of $L(E/K, s)$ around the point 1 in $\mathbb{C}$. We shall see later that the real period of E associated to a global Weierstrass minimal equation appears in the formula for the leading term of this power series expansion. If K has positive class number, then the global Weierstrass minimal equation may not exist (see [1, VIII, Corollary 8.3]). In this case, the real period is defined by integrating a suitably chosen differential on the Néron model of E over K called the Néron differential of E. We will not describe the Néron model of E in this exposition and refer the reader to [10].

The study of L-functions in a broader context was undertaken by Deligne [11]. Deligne identified certain special values of L-functions in this general setting, the so-called 'critical' values of L-functions (at certain integers) and conjectured that these values are algebraic multiples of determinants of matrices whose entries are 'periods'. We merely point to the part of Deligne's conjecture which predicts that in the specific case of the L-function of an elliptic curve defined over $\mathbb{Q}$, the critical value at the integer 1, when divided by a suitable period gives a rational number. Put

$$\mathcal{L}_E = \lim_{s \to 1} L(E, s)/(s-1)^{r_E}.$$

The Birch and Swinnerton-Dyer conjecture (see Conjecture 2c below) predicts an exact formula for the ratio $\mathcal{L}_E/\Omega_E$.

For a nice exposition of the period lattice in the case of elliptic curves, see [11]. For the exact formula in the BSD conjecture, we stress that the choice of the period is important, and will be the real period as in (8). We remark in passing that the theory of periods is profound in itself, a full exposition of which would require delving into cohomology theories and algebraic geometry. For a more in-depth exposition of periods in general, the interested reader is referred to [12].

Height function on Elliptic curves. Fix an algebraic closure $\bar{K}$ of a number field K. For the projective n-space $\mathbb{P}^n$, the absolute *logarithmic height* H_n is the function

$$H_n : \mathbb{P}^n(\bar{K}) \longrightarrow [0, \infty)$$

$$H_n([x_0, \ldots, x_n]) = \sum_{all\ places\ v} (max\{\log |x_0|_v, \ldots, \log |x_n|_v\})$$

where $| - |_v$ is the absolute value at v normalized so that $\prod_v |x|_v = 1$ for all $x \neq 0$ in K. It can be easily checked that H is well defined. For an elliptic curve E defined over K by the Weierstrass equation

$$y^2 = x^3 + ax^2 + bx + c,$$

consider the morphism of projective varieties $f : E(\bar{K}) \longrightarrow \mathbb{P}^1(\bar{K})$ given by $f([x : y : 1]) = [x : 1]$ and $f([0 : 1 : 0]) = [0 : 1]$. The *naïve height* on $E(\bar{K})$ is the function defined as

$$h : E(\bar{K}) \longrightarrow [0, \infty)$$

$$h(P) = H_1(f(P)).$$

Finally, the *canonical height* (also called Néron-Tate height) is the function

$$\hat{h} : E(K) \longrightarrow [0, \infty)$$

$$\hat{h}(P) := \lim_{n\to\infty} 4^{-n} h([2^n]P).$$

Theorem 15 (Néron, Tate) *Let E/K be an elliptic curve and $\hat{h}$ be the canonical height on E.*
(a) $\hat{h}(P) = (1/2)h(P) + O(1)$ for all $P \in E(\bar{K})$.
(b) $\hat{h}(P) = 0$ if and only if P is a torsion point.
(c) The canonical height $\hat{h}$ is a quadratic form on $E(\bar{K})$, i.e. $\hat{h}$ is an even function, and the pairing

$$< \cdot, \cdot >: E(\bar{K}) \times E(\bar{K}) \longrightarrow \mathbb{R}$$

$$< P, Q >= \hat{h}(P + Q) - \hat{h}(P) - \hat{h}(Q)$$

is bilinear.

(d) For all $P \in E(\bar{K})$ *and all* $m \in \mathbb{Z}$*, we have*

$$\hat{h}([m]P) = m^2\hat{h}(P).$$

(e) For all $P, Q \in E(\bar{K})$, $\hat{h}(P+Q) + \hat{h}(P-Q) = 2\hat{h}(P) + 2\hat{h}(Q)$
(f) The canonical height $\hat{h}$ *extends to a positive definite quadratic form on the vector space* $E(K) \otimes \mathbb{R}$.
(d) Any function $E(K) \longrightarrow \mathbb{R}$ *which satisfies (a) and (d) is equal to the canonical height function* $\hat{h}$.

We remark that the above properties of the canonical height function are crucially used in the proof of the Mordell-Weil Theorem (see [1, Chap. VIII]).

Definition 16 The Néron-Tate height pairing on $E(K)$ is the bilinear form

$$< P, Q >_{NT} = (\hat{h}(P+Q) - \hat{h}(P) - \hat{h}(Q)) \text{ for all } P, Q \in E(K).$$

The *elliptic regulator* of E over K is defined by

$$Reg_K(E) := det(< P_i, P_j >_{NT})_{1 \leq i, j \leq r_K(E)}$$

where $P_i, \ldots, P_{r_K(E)}$ is a basis for $E(K)/E(K)_{tor}$.

As a consequence of Theorem 15(f) we have the following

Corollary 17 *The elliptic regulator* $Reg_K(E) > 0$.

Selmer group and Tate-Shafarevich group of elliptic curves. In this section, we discuss the Galois cohomology of elliptic curves and define the Tate-Shafarevich group which is an important invariant associated to an elliptic curve defined over a number field. It is conjectured to be finite and its size appears in the exact formula for the leading term of the power series expansion of the associated Hasse-Weil L-function as predicted by BSD conjectures which is discussed below.

For a field K and a discrete module A over the absolute Galois group $\mathrm{Gal}(\bar{K}/K)$ of K, let $H^i(K, A)$ denote the i-th Galois cohomology group of A. Let E be an elliptic curve defined over K. For an abelian group A let A_{tor} denote the torsion subgroup of A. The absolute Galois group of K acts continuously on the discrete group $E(\bar{K})$ (resp. $E(\bar{K})_{tor}$). Now, suppose that K is a number field and let E be an elliptic curve defined over K. We denote by E_{tor} the Galois module $E(\bar{K})_{tor}$. For every prime v of K, we have a natural restriction map from $H^1(K, E_{tor}) \longrightarrow H^1(K_v, E(\bar{K}_v))$ induced by the inclusion $E_{tor} \hookrightarrow E(\bar{K}_v)$. The Selmer group $\mathrm{Sel}(E/K)$ of E over K is defined as

$$\mathrm{Sel}(E/K) := Ker(H^1(K, E_{tor}) \longrightarrow \prod_v H^1(K_v, E(\bar{K}_v))$$

and the Tate-Shafarevich group denoted by $\text{Ш}(E/K)$, is defined as

$$Ш(E/K) := Ker(H^1(K, E(\bar{K})) \longrightarrow \prod_v H^1(K_v, E(\bar{K}_v))$$

where v varies over the set of primes of K. Using the fact that $E(\bar{K})$ is divisible, for every positive integer m, we get the exact sequence

$$0 \longrightarrow E(\bar{K})[m] \longrightarrow E(\bar{K}) \xrightarrow{\times m} E(\bar{K}) \longrightarrow 0.$$

From the associated long exact sequence of Galois cohomology for every positive integer m, we have the exact sequence

$$0 \longrightarrow E(K)/mE(K) \longrightarrow H^1(K, E(\bar{K})[m]) \longrightarrow H^1(K, E(\bar{K}).$$

Since

$$\varinjlim_m H^1(K, E(\bar{K})[m]) = H^1(K, E(\bar{K})_{tor}),$$

taking the direct limit over integers m, we obtain the exact sequence,

$$0 \longrightarrow E(K) \otimes \mathbb{Q}/\mathbb{Z} \longrightarrow H^1(K, E(\bar{K})_{tor}) \longrightarrow H^1(K, E(\bar{K}).$$

Using this exact sequence along with the snake lemma we get that

$$0 \longrightarrow E(K) \otimes \mathbb{Q}/\mathbb{Z} \longrightarrow \mathrm{Sel}(E/K) \longrightarrow Ш(E/K) \longrightarrow 0$$

is exact. For a prime p, let $\mathrm{Sel}_p(E/K)$ denote the p-primary subgroup of $\mathrm{Sel}(E/K)$.

Proposition 18 *For any number field K, an elliptic curve E defined over K and a prime p, we have an isomorphism*

$$\mathrm{Sel}_p(E/K) \cong (\mathbb{Q}_p/\mathbb{Z}_p)^{s_p(E/K)} \oplus Fin_p,$$

where Fin_p is a finite group.

The number $s_p(E/K)$ is called the *p-Selmer rank of E over K*.

The conjectures of Birch and Swinnerton-Dyer (BSD). We are now ready to state the Birch and Swinnerton-Dyer conjectures as formulated by Birch and Swinnerton-Dyer following extensive numerical calculations that they made in the 1960s (see [13, 14]).

Conjecture 2 (Birch, Swinnerton-Dyer) *For a number field K and an elliptic curve defined over K,*
(a) $order_{s=1} L(E/K, s) = r_K(E)$.
(b) The Tate-Shararevich group $Ш(E/K)$ is finite. In particular, $s_p(E/K) = r_K(E)$ for every prime p.

(c)

$$\lim_{s\longrightarrow 1}\frac{L(E/K,s)}{(s-1)^{r_K(E)}}=\frac{\Omega_{E/K}\times Reg_K(E)\times \#Ш(E/K)\prod_{v<\infty}c_{K_v}(E)}{\sqrt{disc_K}\times(\#E(K)_{tor})^2}$$

where $disc_K$ denotes the discriminant of the field K over $\mathbb{Q}$.

The order of the Hasse-Weil L-series at $s=1$ is called the *analytic rank* of E. Conjecture 2(a) is usually referred to as the weak BSD conjecture and all assertions together as the strong form of *BSD conjecture*. Conjecture 2(c) is also known as the *BSD exact formula.*

The following quote of Tate neatly summarizes the mystery of the BSD conjecture: *"This remarkable conjecture relates the behavior of a function L at a point where it is not at present known to be defined to the order of a group* Ш *not known to be finite".*

The BSD conjectures have been proved in some special cases due to work of Coates-Wiles, Gross–Zagier, Kolyvagin, Rubin and others. We mention that Iwasawa theory and the theory of Euler systems have proved to be effective tools in attacking the BSD conjectures. In 1977, Coates and Wiles proved the following

Theorem 19 *If E is an elliptic curve defined over $\mathbb{Q}$ or a quadratic imaginary extension K of $\mathbb{Q}$, E has complex multiplication by K and $L(E/K,s)$ is non-zero at* 1, *then $E(K)$ is finite.*

The best result known to date about the weak BSD conjecture is the following.

Theorem 20 *If $order_{s=1}L(E/\mathbb{Q},s)\leq 1$, then $order_{s=1}L(E/\mathbb{Q},s)=r_E$ and $Ш(E/\mathbb{Q})$ is finite.*

There are two key ingredients in the proof :
(i) The so-called 'Heegner points' and the deep results of Gross–Zagier [15] relating the canonical height of Heegner points to derivatives of the L-functions. The Gross–Zagier formula implies that if an elliptic curve E over $\mathbb{Q}$ has analytic rank of E equal to one, then $r_E\geq 1$.
(ii) The notion of Euler systems as developed by Kolyvagin [16]. Kolyvagin developed the device of Euler systems, which is a tool that connects the analytic and algebraic sides of the BSD conjecture, and can be used to bound the size of Selmer groups. Kolyvagin used it along with Gross–Zagier's formula to prove the finiteness of $Ш(E/\mathbb{Q})$ and the weak BSD conjecture for elliptic curves over $\mathbb{Q}$ with analytic rank at most one (see [17] for more detail).

For the theory of Euler systems, the interested reader is referred to W. McCallum [18] and to the excellent book by Rubin [19]. The exact formula of the BSD conjecture is known in special cases and uses very different methods, mainly from Iwasawa theory (see [20–23]). For more results on the full BSD conjecture using Iwasawa theory, the reader is referred to the work of Kato [24], Skinner-Urban [25]. Even an outline of these results and the methods therein are beyond the scope of this article.

All the above results assumed the truth of Conjecture 1, which itself was proved by Wiles et al. in the late 1990s (see [7, 9]). There is also a related conjecture known as the *parity conjecture*.

Conjecture 3 $order_{s=1}L(E/K, s) = r_K(E) \bmod 2.$

Note that the parity conjecture is a consequence of BSD conjecture. It is still open in general. In recent years, there has been some significant progress towards the proof another related conjecture, namely the p-$parity$ conjecture due to Greenberg, Nekovar, Dokchitser-Dokschitser, Wei Zhang and others (see [26–29]).

Conjecture 4 (p-parity conjecture) $order_{s=1}L(E/K, s) = s_p(E/K) \bmod 2$ *for all primes* p.

The p-parity conjecture together with finiteness of the p-torsion subgroup of Shafarevich-Tate group implies the parity conjecture. Monsky proved the following

Theorem 21 ([30]) *(a) The 2-parity conjecture is true for an elliptic curve E defined over* $\mathbb{Q}$.
(b) The p-parity conjecture is true for an elliptic curve E defined over $\mathbb{Q}$ *if it has a p-isogeny defined over* $\mathbb{Q}$.

Due to work of Nekovar and Dokchitser-Dokschister, the p-parity conjecture is known in general over $\mathbb{Q}$. Over a more general totally real number field, we have the following theorem by Nekovar.

Theorem 22 ([28]) *Let E be an elliptic curve defined over a totally real field F. Then the p-parity conjecture holds for E over F in each of the following situations*
(i) E does not have complex multiplication
(ii) E has complex multiplication and $2 \nmid [F : \mathbb{Q}]$
(iii) E has complex multiplication by an immaginary quadratic field K' *and p splits completely in* $K'/\mathbb{Q}$.

For more precise technical results on the p-parity conjecture over some other number fields by Dokchitser-Dokchitser see [31].

We close with a brief mention of what is known about the BSD conjecture when one considers all elliptic curves over number fields. Even though general results on BSD are not known for elliptic curves of algebraic rank > 1, the conjecture is known to hold for a large class of curves. Manjul Bhargava and Arul Shankar introduce and study the '*average rank*' of elliptic curves and show that the average rank is less than one (see [32, 33]). Recent work of Bhargava, Skinner and Zhang implies that 66% of the class of all elliptic curves satisfy the BSD conjecture (see [34]) . In fact, Bhargava and Shankar show that in a statistical sense, a sizeable proportion of elliptic curves defined over $\mathbb{Q}$ has rank zero and another sizeable proportion has rank one (see [34, Theorem 3] for a precise statement).

So far, there is no general algorithm known which can compute the rank of a given elliptic curve defined over a number field. One can compute the rank of an elliptic curve by computing the derivative of the associated Hasse-Weil L-function only if BSD is known. But at present, we do not know a single example of an elliptic curve over $\mathbb{Q}$ of rank ≥ 2 for which $\text{III}(E/\mathbb{Q})$ is finite.

Numerical examples. The BSD conjectures have been verified extensively for numerous concrete examples. As remarked earlier, the conjecture itself was formulated based on the data obtained by explicit computations. In the last half a century, remarkable progress on the computational side has been made possible, thanks to advances in computing and theoretical knowledge. We refer the reader to the excellent data base compiled by Cremona, Stein, Watkins and others [35, 36]. Thus the beauty of BSD conjectures lies in its intricacy combined with the fact that most of the invariants in the exact formula can be explicitly computed.

In this final section, we provide a few illustrative numerical examples using the mathematical software Sage. We shall provide three examples of elliptic curves with analytic rank zero, one and two respectively. Consider the ellipic curve E given by the Weierstrass equation

$$E : y^2 + y = x^3 - x^2 - 7820x - 263580$$

over $\mathbb{Q}$. The discriminant Δ of E over $\mathbb{Q}$ is 11 and the j invariant of E is equal to $1 \times 212 \times 11^{-1} \times 29^3 \times 809^3$. The curve E has split multiplicative reduction at 11. The BSD invariants of E are as follows:

- analytic rank, $r = 0$,
- regulator, $Reg_{\mathbb{Q}}(E) = 1$,
- real period $\Omega = 0.253841860856\ldots$
- torsion order, $\#E(\mathbb{Q})_{tor} = 1$
- $L(E/\mathbb{Q}, 1) = 0.253841860856\ldots$
- Tamagawa Number at 11, $c_{11}(E) = 1$.

Since $L(E/\mathbb{Q}, 1) \neq 0$, BSD conjectures are true for E over $\mathbb{Q}$. In particular, $r_{\mathbb{Q}}(E) = 0$ and from the above data we get that $\#\text{III}(E/\mathbb{Q}) = 1$. Next, we consider the following elliptic curve E of positive rank :

$$E : y^2 + y = x^3 + x^2.$$

The curve E has non-split multiplicative reduction at 43. The discriminant of E is -43 and the j invariant is $-1 \times 3^{12} \times 43^{-1}$. The BSD invariants of E are given by

- analytic rank, $r = 1$,
- regulator, $Reg_{\mathbb{Q}}(E) = 0.0628165070875\ldots$,
- real period $\Omega = 5.46868952997\ldots$
- torsion order, $\#E(\mathbb{Q})_{tor} = 1$
- $L(E/\mathbb{Q}, 1) = 0$ and $L'(E/\mathbb{Q}, 1) = 0.343523974618\ldots$
- Tamagawa Number at 43, $c_{43}(E) = 1$.

Since E has analytic rank 1 over $\mathbb{Q}$, BSD conjectures hold and we get that $r_{\mathbb{Q}}(E) = 1$ and $\#\text{Ш}(E/\mathbb{Q}) = 1$. We shall end by providing an example of elliptic curve for which we still do not know if the BSD conjectures are true. Consider the elliptic curve

$$E : y^2 + y = x^3 + x^2 - 2x.$$

The curve E has split multiplicative reduction at 389. The discriminant of E is 389 and the j invariant is $2^{12} \times 7^3 \times 389^{-1}$. The BSD invariant of E are given by

- analytic rank, $r = 2$,
- regulator, $Reg_{\mathbb{Q}}(E) = 0.15246017794\ldots$,
- real period $\Omega = 4.98042512171\ldots$
- torsion order, $\#E(\mathbb{Q})_{tor} = 1$
- $L(E/\mathbb{Q}, 1) = 0$ and $L'(E/\mathbb{Q}, 1) = 0$ and $L''(E/\mathbb{Q}, 1) = 0.759316500288\ldots$
- Tamagawa Number at 389, $c_{389}(E) = 1$.

The BSD conjecture for E over $\mathbb{Q}$ predicts that the $r_{\mathbb{Q}}(E) = 2$ and $\#\text{Ш}(E/\mathbb{Q}) = 1$. In this case it can be shown (using the method of "2-descent") that $r_{\mathbb{Q}}(E) = 2$ and $\#\text{Ш}(E/\mathbb{Q})[2] = 1$ (see [37]). A set of generators of the Mordell-Weil group of E over $\mathbb{Q}$ is given by $\{(1, 1), (0, 0)\}$. At present we do not know if the predictions on the size of $\#\text{Ш}(E/\mathbb{Q})$ is true.

References

1. J.H. Silverman, *The Arithmetic of Elliptic Curves*, 2nd edn. Graduate Texts in Mathematics, vol. 106 (Springer, Berlin, 2009)
2. H. Darmon, Rational points on modular elliptic curves, http://www.math.mcgill.ca/darmon/pub/Articles/Research/36.NSF-CBMS/chapter.pdf
3. A.J. Wiles, The Birch and Swinnerton-Dyer conjecture, http://www.claymath.org/prizeproblems/birchsd.htm
4. J.H. Silverman, J. Tate, *Rational Points on Elliptic Curves*. Undergraduate Text in Mathematics (Springer, Berlin, 1992)
5. J. Tate, *A Review of Non-Archimedean Elliptic Functions, Elliptic Curves, Modular Forms and Fermats Last Theorem* (International Press, Somerville, 1995), pp. 162–184
6. J.H. Silverman, *Advanced Topics in the Arithmetic of Elliptic Curves*, vol. 151. Graduate Texts in Mathematics (Springer, Berlin, 1994)
7. C. Breuil, B. Conrad, F. Diamond, R. Taylor, On the modularity of elliptic curves over Q: wild 3-adic exercises. J. Am. Math. Soc. **14**(4), 843939 (electronic). MR 2002d:11058 (2001)
8. L. Clozel, Base change for $GL(n)$, in *Proceedings of the International Congress of Mathematicians* (Berkeley California, USA, 1986)
9. A.J. Wiles, Modular elliptic curves and Fermats last theorem. Ann. Math. (2) **141**(3), 443551. MR 1333035 (96d:11071) (1995)
10. S. Bosch, W. Lutkeböhmert, N. Raynaud, *Néron Models*. A Series of Modern Serveys in Mathematics (Springer, Berlin, 1989)
11. O. Debarre, Period of algebraic varieties, http://www.math.ens.fr/~debarre/ExposeLille.pdf
12. M. Kontsevich, D. Zagier, Periods, http://www.maths.ed.ac.uk/~aar/papers/kontzagi.pdf
13. B.J. Birch, *Elliptic Curves Over Q: A Progress Report, 1969 Number Theory Institute (Proceedings of Symposium Pure Mathematics, vol. xx, State University, New York, Stony Brook, 1969)* (American Mathematical Society, Providence, 1971), pp. 396–400

14. B. Birch, H. Swinnerton-Dyer, Notes on elliptic curves II. Journ. Reine u. Angewandte Math. **218**, 79–108 (1965)
15. B.H. Gross, D.B. Zagier, Heegner points and derivatives of L-series. Invent. Math. **84**, 225–320 (1986)
16. V. Kolyvagin, Finiteness of $E(Q)$ and Ш(E/Q) for a class of Weil curves. Izv. Akad. Nauk SSSR **52**, 523–541 (1988). Translation Math. USSR-Izv. **32** (1989)
17. W. Zhang, *The Birch-Swinnerton-Dyer Conjecture and Heegner Points: A Survey. Current Developments in Mathematics 2013* (International Press, Somerville, 2014), pp. 169–203
18. W. McCallum, *Kolyvagin's Work on Shafarevich-Tate Groups. L- Functions and Arithmetic (Durham, 1989)*, London Mathematical Society, Lecture Note Series, vol. 153 (Cambridge University Press, Cambridge, 1991), pp. 295–316
19. K. Rubin, *Euler Systems*, vol. 147. Annals of Mathematics Studies (Princeton University Press, Princeton, 2000)
20. J. Coates, A. Wiles, On the conjecture of Birch and Swinnerton-Dyer. Invent. Math. **39**, 233–251 (1977)
21. K. Rubin, The main conjectures of Iwasawa theory for imaginary quadratic fields. Invent. Math. **103**, 25–68 (1991)
22. K. Rubin, Tate-Shafarevich groups and L-functions of elliptic curves with complex multiplication. Invent. Math. **89**, 527–560 (1987)
23. K. Rubin, On the main conjecture of Iwasawa theory for imaginary quadratic fields. Invent. Math. **93**, 701–713 (1988)
24. K. Kato, p-adic Hodge theory and values of zeta functions of modular forms. Cohomologies p-adiques et applications arithmtiques III. Astérisque **295**, ix, 117–290 (2004)
25. C. Skinner, E. Urban, The Iwawasa main conjectures for GL2. Invent. Math. **195**(1), 1–277 (2014)
26. T. Dokchitser, V. Dokchitser, Root numbers and parity of ranks of elliptic curves. J. Reine Angew. Math. **658**, 39–64 (2011)
27. R. Greenberg, *Iwasawa Theory, Projective Modules, and Modular Representations*, vol. 211. Memoirs of the American Mathematical Society, vol. 992 (American Mathematical Society, Providence, 2011)
28. J. Nekovar, On the parity of ranks of Selmer groups IV. Compositio Math. **145**(6), 1351–1359 (2009)
29. W. Zhang, Selmer groups and the indivisibility of Heegner points. Camb. J. Math. **2**(2), 191–253 (2014)
30. P. Monsky, Generalizing the Birch-Stephens theorem I modular curves. Math. Z. **221**(3), 415–420 (1996)
31. Notes on the parity conjecture, *September 2010, CRM Barcelona Advanced Courses in Mathematics Elliptic Curves, Hilbert Modular Forms and Galois Deformations* (Birkhauser, Basel, 2013)
32. M. Bhargava, A. Shankar, Binary quartic forms having bounded invariants, and the boundedness of the average rank of elliptic curves. Ann. Math. (2) **181**(1), 191–242 (2015)
33. M. Bhargava, A. Shankar, Ternary cubic forms having bounded invariants, and the existence of a positive proportion of elliptic curves having rank 0. Ann. Math. (2) **181**(2), 587–621 (2015)
34. M. Bhargava, C. Skinner, W. Zhang, A majority of elliptic curves over $\mathbb{Q}$ satisfy the Birch and Swinnerton-Dyer conjecture, arxiv:407.1826v2.pdf
35. J.E. Cremona, *Algorithms for Modular Elliptic Curves*, 2nd edn. (Cambridge University Press, Cambridge, 1997), http://www.maths.nott.ac.uk/personal/jec/book/
36. W.A. Stein, Modular forms database, http://modular.math.washington.edu/Tables/
37. J.E. Cremona, Numerical evidence for the Birch and Swinnerton-Dyer conjecture, http://homepages.warwick.ac.uk/staff/J.E.Cremona/papers/bsd50.pdf

Kolyvagin's Work on Modular Elliptic Curves

J.-J. Lee

Abstract This exposition is a short introduction to the theory of Euler systems by Kolyvagin. We construct explicit cohomology classes coming from Heegner points, and outline the proof of the Birch and Swinnerton-Dyer conjecture when the order of vanishing of L-function attached to an elliptic curve is either 0 or 1. We define an Euler system by making the essential properties in this construction the axioms of the system.

1 Introduction

Let G be a group and A, B, C be G-modules such that

$$0 \to A \to B \to C \to 0$$

is an exact sequence. Then there corresponds a long exact sequence of cohomology groups

$$0 \to A^G \to B^G \to C^G \to H^1(G, A) \to H^1(G, B) \to \cdots,$$

where $A^G := \{a \in A \mid ga = a \text{ for } \forall g \in G\}$ and

$$H^q(G, A) := (q - \text{cocycles})/(q - \text{coboundaries}).$$

Especially, 1-cocycle is a map $\varphi : G \to A$ such that $\varphi(g_1 g_2) = g_1\varphi(g_2) + \varphi(g_1)$ and 1-coboundary is a 1-cocycle φ such that there is an element a in A that satisfies $\varphi(g) = ga - a$ for all $g \in G$.

We will consider elliptic curves as G-modules where G is a Galois group.

Let E be an elliptic curve defined over a number field K. In other words, E is given by a Weierstrass equation

J.-J. Lee (✉)
Gyeonguiro 333, Ilsan, Goyang 10417, South Korea
e-mail: jungjolee@gmail.com

Z. Liang and C. Aribam (eds.), *The Computational and Theoretical Aspects of Elliptic Curves*, Mathematical Lectures from Peking University,
https://doi.org/10.1007/978-981-13-6664-2_2

$$E : y^2 = x^3 + ax + b, \quad \text{where } a, b \in K$$

with its discriminant $\Delta = -16(4a^3 + 27b^2) \neq 0$.

Let L/K be a Galois extension. Then $E(L)$ is a $G(L/K)$-module under the action

$$\begin{array}{rcl} G(L/K) \times E(L) & \to & E(L) \\ (g \ , \ (x, y)\ ) & \mapsto & (gx, gy). \end{array}$$

For a natural number M, let E_M denote

$$E_M = E(\overline{K})_M := \{P \in E(\overline{K}) \mid MP = O\},$$

where $\overline{K}$ is the algebraic closure of K. It is a $G(\overline{K}/K)$-module.

Let's write $G_K = G(\overline{K}/K)$. We have a short exact sequence of G_K-modules

$$O \to E_M \to E(\overline{K}) \overset{\times M}{\to} E(\overline{K}) \to O.$$

From the corresponding long exact sequence, we get

$$O \to E(K)/ME(K) \to H^1(G_K, E_M) \to H^1(G_K, E(\overline{K}))_M \to O,$$

called the Kummer exact sequence.

If F is a field and A is an abelian group, we shall abbreviate $H^i(G(\overline{F}/F), A)$ as $H^i(F, A)$, where A is a $G(\overline{F}/F)$-module, for $i \in \mathbb{N}$. For a place v of a number field K, we let $K(v)$ denote the corresponding completion of K. Consider a series of maps

$$\begin{aligned} H^1(K(v), E_M) \times H^1(K(v), E_M) &\overset{(a)}{\to} H^2(K(v), E_M \otimes E_M) \\ &\overset{(b)}{\to} H^2(K(v), \mu_M) \overset{(c)}{\to} H^2(K(v), \overline{K(v)}^{\times})_M \overset{(d)}{\to} \mathbb{Z}/M\mathbb{Z}. \end{aligned}$$

The map (a) is a cup product map defined on cohomology groups (see [1], p. 105). The map (b) is induced by the Weil pairing $[\ ,\ ]_M : E_M \otimes E_M \to \mu_M$. The map (c) is an isomorphism obtained by applying the fact that $H^1(K(v), \overline{K(v)}^{\times}) = 0$ (Hilbert 90) on the long exact sequence corresponding to $1 \to \mu_M \to \overline{K(v)}^{\times} \overset{\nu}{\to} \overline{K(v)}^{\times} \to 1$, where ν is a map given by $\nu : x \mapsto x^M$ and $G(\overline{K(v)}/K(v))$ acting on them. Finally, the map (d) is the invariant map defined on the Brauer group $H^2(K(v), \overline{K(v)}^{\times})$ (see [1], Chap. VI, §1).

Composing the maps from (a) to (d) above, we obtain a bilinear paring

$$\langle\ ,\ \rangle_{v,M} : H^1(K(v), E_M) \times H^1(K(v), E_M) \to \mathbb{Z}/M\mathbb{Z} \tag{1}$$

which is called the Tate pairing.

The group $E(K(v))/M = E(K(v))/ME(K(v))$ is self-dual under Tate pairing. Indeed, if we define

$$\big(E(K(v))/M\big)^{\perp} := \big\{x \in H^1(K(v), E_M) \mid \langle e, x\rangle_{v,M} = 0 \quad \forall e \in E(K(v))/M\big\}$$

by identifying $E(K(v))/ME(K(v))$ as a subgroup of $H^1(K(v), E_M)$ via the Kummer exact sequence, then the self-duality says that

$$\big(E(K(v))/ME(K(v))\big)^{\perp} = E(K(v))/ME(K(v)).$$

From this property, we obtain a "non-degenerate" bilinear pairing

$$\langle\ ,\ \rangle_{v,M} : E(K(v))/ME(K(v)) \times H^1(K(v), E)_M \to \mathbb{Z}/M\mathbb{Z} \tag{2}$$

which we also call the (local) Tate pairing (at v) with the same notation as in (1).

We can reformulate the global reciprocity law in the following way.

Theorem 1 *If $\alpha \in Br(K)$, then $\sum_{v\in\mathfrak{M}_K} inv_v(\alpha) = 0$, where $\mathfrak{M}_K$ denotes the set of places of K. (Here, $Br(K)$ denotes $H^2(K, \overline{K}^{\times})$ by definition.)*

Proof This is Theorem B, Chap. VII of [1].

Let $D(v) = G(\overline{K(v)}/K(v))$. It is a subgroup of $G(\overline{K}/K)$, which gives us a restriction homomorphism $H^1(G(\overline{K}/K), E(\overline{K})) \to H^1(D(v), E(\overline{K}))$. Composing it with the homomorphism $H^1(D(v), E(\overline{K})) \to H^1(D(v), E(\overline{K(v)}))$ derived from $E(\overline{K}) \subset E(\overline{K(v)})$, we get the localization map (at v)

$$\begin{array}{ccc} H^1(G(\overline{K}/K), E(\overline{K})) & \to & H^1(G(\overline{K(v)}/K(v)), E(\overline{K(v)})). \\ \varphi & \mapsto & \varphi_v \end{array}$$

Definition 2 (*Selmer group*) The Selmer group $S(K, E_M)$ of E/K of level M is defined to be

$$S(K, E_M) := \{x \in H^1(K, E_M) \mid x_v \in E(K(v))/ME(K(v)) \ \text{ for every } v \in \mathfrak{M}_K\}$$

where x_v is the localization of x at v.

Therefore, in our situation, Theorem 1 can be reinterpreted as follows.

Theorem 3 (Orthogonality Relation) *For $P \in S(K, E_M)$ and $C \in H^1(K, E)_M$, we have*

$$\sum_v \langle P_v, C_v\rangle_{v,M} = 0$$

where P_v and C_v are the localizations of P and C respectively at each $v \in \mathfrak{M}_K$.

2 Explicit Cohomology Classes

Let G be a cyclic group of finite order. For a G-module A, we can describe its cohomology groups in terms of the generator of the group G. Let σ be a generator of G of finite order. Let N denote the norm map on A defined by

$$\begin{aligned} N \ :A &\to \quad A \\ a &\mapsto \textstyle\sum_{\sigma \in G} \sigma a. \end{aligned}$$

Then

$$H^q(G, A) = \begin{cases} A^G/NA, & \text{if } q \geqslant 2 \text{ and even;} \\ \ker(N : A \to A)/(\sigma - 1)A, & \text{if } q \text{ is odd.} \end{cases}$$

Reference ([1], p. 108).

We will apply this fact in our situation on elliptic curves.

Let E be an elliptic curve defined over a number field F and L/F a Galois extension with group $G_{L/F} = \langle \sigma \rangle$. Then $E(L)$ is a $G_{L/F}$-module and

$$H^1(G_{L/F}, E(L)) \simeq \{P \in E(L) \mid \mathrm{Norm}_{L/F}\, P = O\}/(\sigma - 1)E(L). \tag{3}$$

Here, $\mathrm{Norm}_{L/F} : E(L) \to E(F)$ is defined by $P \mapsto \sum_{g \in G_{L/F}} gP$.

In view of (3), if we can find cyclic extensions L/F of finite degree with points of trivial norm on such extensions where F is an extension of K, it will be possible to describe certain explicit elements in $H^1(G_{L/F}, E(L))$, and hence those in $H^1(K, E)_M$, where $M = [L : F]$. These elements can be used to deduce information on $S(K, E_M)$ via the orthogonality relation described in Theorem 3. An answer to this question given by Kolyvagin was that there are subextensions of certain ring class fields that are cyclic, where the norm relation between Heegner points gives us a formula for finding points of trivial norm on such extensions.

Let E be an elliptic curve with conductor N. Let K be an imaginary quadratic field with discriminant D such that

$$D \equiv (\text{a square}) \pmod{4N}, \quad D \not\equiv -3, -4, \quad (D, 2N) = 1. \tag{4}$$

Let O_K be the ring of integers of K. For a positive integer λ, let $O_\lambda = \mathbb{Z} + \lambda O_K$ be an order of K of conductor λ, that is, it is the unique subring of O_K of index λ.

In general, for an order O in K, $j(O)$ is an algebraic integer, where j is the j-invariant considering O as a lattice.

The field $K_\lambda = K(j(O_\lambda))$ is called the ring class field of conductor λ. It is an algebraic extension of K. In particular, K_1 is called the Hilbert class field of K.

Class field theory says that there is a canonical isomorphism

$$G(K_\lambda/K_1) \cong (\mathcal{O}_K/\lambda\mathcal{O}_K)^\times/(\mathbb{Z}/\lambda\mathbb{Z})^\times.$$

In particular, for an odd prime p that is unramified in $K/\mathbb{Q}$,

$$G(K_p/K_1) \text{ is cyclic of order } p - \left(\frac{p}{K}\right) = \begin{cases} p+1, & \text{if } p \text{ is inert in } K; \\ p-1, & \text{if } p \text{ splits in } K. \end{cases}$$

(In fact, K_p/K_1 is totally ramified as well.)

From now on, we will assume that $\left(\frac{p}{K}\right) = -1$, because considering this case is enough to prove the result of Kolyvagin on the Birch and Swinnerton-Dyer conjecture.

Let $X_0(N)(\mathbb{C}) = \Gamma_0(N) \setminus (\mathcal{H} \cup \mathbb{P}^1(\mathbb{Q}))$, where $\mathcal{H}$ is the complex upper half plane. Then there is a one-to-one correspondence between $X_0(N)(\mathbb{C}) \setminus \{\text{cusps}\}$ and the set of equivalence classes of isogenies of elliptic curves with cyclic kernel of order N.

Let α be an ideal in O_K such that $O_K/\alpha \simeq \mathbb{Z}/N\mathbb{Z}$. Such α exists because of our condition on $D : 0 > D \equiv (\text{a square}) \pmod{4N}$. In fact, there exist $2^s \cdot h$ of them, where s is the number of distinct prime factors of N and h is the ideal class number of K.

Define $\alpha_\lambda = \alpha \cap O_\lambda$. For $(\lambda, N) = 1$, let $\alpha_\lambda^{-1} = \{x \in K | x\alpha_\lambda \subset O_\lambda\}$ be the inverse of α_λ in the group of proper O_λ-ideals.

Let's denote by $\mathcal{Z}_\lambda$ the point in $X_0(N)(\mathbb{C}) \setminus \{\text{cusps}\}$ that corresponds to the equivalence class of the isogeny $(\mathbb{C}/O_\lambda \to \mathbb{C}/\alpha_\lambda^{-1})$. It is known that $\mathcal{Z}_\lambda \in X_0(N)(K_\lambda)$ and there is a modular parametrization $\gamma : X_0(N)(K_\lambda) \to E(K_\lambda)$. The point $\mathcal{Y}_\lambda = \gamma(\mathcal{Z}_\lambda) \in E(K_\lambda)$ is called the Heegner point of conductor λ for K.

Theorem 4 (Norm relation between Heegner points) *For* $p|\lambda$,

$$\mathrm{Norm}_{K_\lambda/K_{\lambda/p}} \mathcal{Y}_\lambda = a_p \mathcal{Y}_{\lambda/p}$$

where $a_p = p + 1 - [\widetilde{E}(\mathbb{Z}/p\mathbb{Z})]$.

Proof Let $f(\tau) = \sum a_n q^n$ (for $q = e^{2\pi i \tau}$) be the generating function with its coefficients coming from those of $L(E, s)$, the L-series of E. Assume that it is normalized so that $a_1 = 1$. The proof uses the fact that f is a cusp form of weight 2 for $\Gamma_0(N)$ that is an eigenform for the Hecke operator T_p with its corresponding eigenvalue a_p, for each p.

Definition 5 Let M be a natural number such that $M|p+1$ and $M|a_p$. Let L be a subextension of K_p/K_1 with $[L : K_1] = M$. Define $P_L \in E(L)$ by

$$P_L = \mathrm{Norm}_{K_p/L} \mathcal{Y}_p - \frac{a_p}{M} \mathcal{Y}_1.$$

Remark 6 For a given integer M, there exists a prime p satisfying the conditions in the above definition (in fact infinitely many such primes) due to the Chebotarev density theorem. The corresponding Galois extension L of K_1 in K_p exists by Galois theory, since K_p/K_1 is a cyclic extension, all of its subgroups are normal.

When we define P_L as in Definition 5, it is easy to check that $\mathrm{Norm}_{L/K_1}(P_L) = O$.

Theorem 7 (Gross–Zagier formula)

$$L'(K, E, 1) = \beta \cdot \mathrm{height}(P_K)$$

where β is a non-zero real constant determined by E, K and $P_K = \mathrm{Norm}_{K_1/K}\, \mathcal{Y}_1 \in E(K)$.

Waldspurger's theorem is a result which identifies Fourier coefficients of modular forms of half-integral weight $k + 1/2$ with the value of a L-series at $s = k/2$. It implies the following.

Theorem 8 *There exists D satisfying the condition (4) such that $L'(K, E, 1) \neq 0$.*

Theorems 7 and 8 imply that there exists D such that $\mathrm{height}(P_K) \neq 0$, that is, the Heegner point P_K has an infinite order. Therefore, for such D, rank $E(K) \geqslant 1$.

The descent exact sequence

$$O \to E(K)/ME(K) \to S_M \to Ш(K, E)_M \to O,$$

where S_M denotes $S(K, E_M)$, implies that

$$[E(K)/ME(K)] = [(\mathbb{Z}/M\mathbb{Z})^r] = M^r \leqslant [S_M], \tag{5}$$

for any prime $\ell \nmid [E(K)_{\mathrm{tor}}]$, $M = \ell^n$ and $r = \mathrm{rank}\, E(K)$.

By controlling the local behavior of the cohomology classes constructed from Heegner points, one can bound the size of the Selmer group as

$$[S_M] \leqslant M\ell^{2m_0}$$

where m_0 comes from an annihilating ideal of certain elements of an Euler system.

Take large enough n, say $n > 2m_0$. Equation (5) implies that $\ell^{nr} \leqslant \ell^{n+2m_0}$, that is, $nr \leqslant n + 2m_0$. On the other hand, our choice of K implies that $r \geqslant 1$, because the Heegner point $P_K \in E(K)$ has an infinite order. Thus, we conclude that $r = 1$.

What we just proved is the following.

Theorem 9 *If the Heegner point has infinite order, then* rank $E(K) = 1$. *Moreover,* $[Ш(K, E)_{\ell^\infty}] \leqslant \ell^{2m_0}$

Let $C_K = [E(K)/\mathbb{Z}P_K]$. One can prove that $m_0 = \mathrm{ord}_\ell\, C_K$. (See [6], §3.)

Let σ be the generator of $G(K/\mathbb{Q})$, that is, $G(K/\mathbb{Q}) = \langle 1, \sigma \rangle$. Let $[E(K)_{\mathrm{tor}}] = m$. Then a theorem of Manin–Drinfeld (later by Mazur) implies that

$$mP_K^\sigma = (-1)^{r_{an}-1} mP_K \tag{6}$$

where $r_{an} = \mathrm{ord}_{s=1}\, L(E, s)$.

Lemma 10 *Let $K = \mathbb{Q}(\sqrt{D})$. An elliptic curve E is given by $E : y^2 = f(x)$, and E_D by $E_D : Dy^2 = f(x)$. Then*

$$\text{rank } E(K) = \text{rank } E(\mathbb{Q}) + \text{rank } E_D(\mathbb{Q}).$$

Proof See [8], Chap. X, Exercise 10.22.

Thus, rank $E(\mathbb{Q})$ + rank $E_D(\mathbb{Q}) = 1$ if the Heegner point has an infinite order.

Let's assume $r_{an} \leqslant 1$.

Suppose $r_{an} = 0$. Then (6) implies that $mP_K^\sigma = -mP_K$, which enables us to conclude that $mP_K \in E_D(\mathbb{Q})$. Therefore, rank $E_D(\mathbb{Q}) = 1$ and rank $E(\mathbb{Q}) = 0$.

Suppose $r_{an} = 1$. Then (6) implies that $mP_K^\sigma = mP_K$, which enables us to conclude that $mP_K \in E(\mathbb{Q})$. Therefore, rank $E(\mathbb{Q}) = 1$ and rank $E_D(\mathbb{Q}) = 0$.

Notice that r_{an} = rank $E(\mathbb{Q})$ in both cases.

3 Norm Relation and Definition of an Euler System

In the previous section, we used Heegner points to construct certain explicit cohomology classes. We now take a different perspective and make the *essential* features used in the construction the axioms of an Euler system.

Let E be an elliptic curve defined over $\mathbb{Q}$ with conductor N. Let K be an imaginary quadratic extension of $\mathbb{Q}$ such that $K \neq \mathbb{Q}(\sqrt{-1}), \mathbb{Q}(\sqrt{-3})$. Let

$$\mathcal{P}(n) := \{p \in \mathbb{N} \mid p \text{ is a prime such that } (p, n) = 1\}.$$

Let K_λ be the ring class field of conductor $\lambda \in \mathbb{Z}$, as defined in Sect. 2.

For $p \in \mathcal{P}(N)$, let

$$m_p := \text{ord}_\ell\left(\gcd\left(p - \left(\frac{p}{K}\right), a_p - 1 - \left(\frac{p}{K}\right)\right)\right).$$

If we let $M = \ell^{m_p}$ for a fixed rational prime ℓ, it follows that

$$M \middle| p - \left(\frac{p}{K}\right) \text{ and } M \middle| a_p - 1 - \left(\frac{p}{K}\right). \tag{7}$$

For $k \in \mathbb{N}$, let Σ be a subset of the set of ordered $(k+1)$-tuples of natural numbers

$$\Sigma \subset \{\lambda \in \mathbb{N} \mid (\lambda, N) = 1\}, \qquad \text{for } k = 0,$$
$$\Sigma \subset \{[\lambda, \pi] \in \mathbb{N}^{k+1} \mid (\lambda, N) = 1, \pi = [\lambda_1, \cdots, \lambda_k] \in \mathcal{P}(N)^k\}, \qquad \text{for } k > 0.$$

For $n_1, n_2 \in \mathbb{N}$, $n_1 \leqslant n_2$, let $M_j = \ell^{n_j}$. Then the homomorphism

$$E_{M_2} \to E_{M_1}$$

given by $P \mapsto \dfrac{M_2}{M_1} P$ induces a homomorphism

$$H^1(K_\lambda, E_{M_2}) \to H^1(K_\lambda, E_{M_1}). \tag{8}$$

We define a set T of cohomology classes as follows:

$$T := \left\{\tau_\lambda = \{\tau_\lambda(\ell^n)\} \in \varprojlim_n H^1(K_\lambda, E_{\ell^n}) \mid \lambda \in \Sigma\right\} \text{ if } k = 0$$

where the projective limit is taken with respect to the maps (8), and

$$T := \left\{\tau_{\lambda,\pi} = \tau_{\lambda,\pi}(M_{\lambda_1}) \in H^1(K_\lambda, E_{M_{\lambda_1}}) \mid [\lambda, \pi] \in \Sigma\right\} \text{ if } k > 0.$$

Definition 11 (*Corestriction*) Let H be a subgroup of G. Let A, B, C be G-modules such that the sequence $0 \to A \to B \to C \to 0$ is exact. Let $G/H = \cup_i s_i H$. For any $a \in A^H$, define

$$N(a) = \mathrm{Norm}_{G/H}(a) := \sum_i s_i a \in A^G.$$

(It is independent of the choice of coset representatives and fixed by G.)

Then, $\mathrm{Cor} : H^q(H, A) \to H^q(G, A)$ is defined by the following commuting diagram:

$$\begin{array}{ccccccccc}
0 & \longrightarrow & A^H & \longrightarrow & B^H & \longrightarrow & C^H & \longrightarrow & H^1(H, A) & \longrightarrow & \cdots \\
 & & {\scriptstyle N}\downarrow & & \downarrow{\scriptstyle N} & & \downarrow{\scriptstyle N} & & \downarrow{\scriptstyle \exists!\,\mathrm{Cor}} & & \\
0 & \longrightarrow & A^G & \longrightarrow & B^G & \longrightarrow & C^G & \longrightarrow & H^1(G, A) & \longrightarrow & \cdots
\end{array}$$

where the two horizontal sequences are the long exact sequences corresponding to the short exact sequence $0 \to A \to B \to C \to 0$. In other words, the corestriction map $\mathrm{Cor} : H^q(H, A) \to H^q(G, A)$ is obtained from the Norm map by the dimension shifting. Actually, corestrictions are defined for the Tate cohomology groups. (See [1], p. 104.)

Norm Relation. For all $p \in \mathcal{P}(2N\lambda)$ and $[\lambda, \pi] \in \Sigma$ such that the prime divisor $\wp$ of p in K is unramified in K_λ and $[p\lambda, \pi] \in \Sigma$. Then

$$\mathrm{Cor}_{K_{p\lambda}/K_\lambda}(\tau_{p\lambda,\pi}) = \mathcal{L}_p \tau_{\lambda,\pi}$$

where

$$\mathcal{L}_p = \begin{cases} a_p, & \text{if } p \text{ is inert in } K; \\ a_p - \mathrm{Fr}_{\wp_1} - \mathrm{Fr}_{\wp_2}, & \text{if } (p) = \wp_1\wp_2 \text{ in } K. \end{cases}$$

Here, $\mathrm{Fr}_\wp$ is the image of $\wp$ in $G(K_\lambda/K)$ under the Artin map.

Definition 12 The set T defined above is called a k-th order Euler system with index set Σ if it satisfies the Norm Relation.

Consider the L-function

$$L(E, s) = \prod_{p \nmid N} L_p(p^{-s})^{-1} \cdot (\text{a finite product})$$

where $L_p(X) = 1 - a_p X + pX^2$. There is a certain relationship between $\mathcal{L}_p \in \mathbb{Z}[\mathrm{G}(K_\lambda/K)]$ and $L_p^*(\mathrm{Fr}_\wp)$, where $L_p^*(X) = X^2 - a_p X + p$ and $\wp$ is a prime divisor of p in K as given in the Norm Relation above. In this sense, an Euler system can be considered as an algebraic counterpart of the L-function expressed as an Euler product.

4 Derived Systems

Suppose we are given a k-th order Euler system (T, Σ). In this section, we use it to construct a $(k+1)$-th order Euler system (T', Σ').

If $(\lambda, p) = 1$, then we have

$$\mathrm{G}(K_{p\lambda}/K_\lambda) \simeq \mathrm{G}(K_p/K_1) \tag{9}$$

and $\mathrm{G}(K_p/K_1)$ is a cyclic group of order $p - \left(\frac{p}{K}\right)$.

Fix a generator $t_p \in \mathrm{G}(K_p/K_1)$ and regard it as a generator of $\mathrm{G}(K_{p\lambda}/K_\lambda)$ by the identification (9).

Suppose $\lambda_0 \in \mathcal{P}(2N\lambda)$ such that $[\lambda, \pi], [\lambda_0\lambda, \pi] \in \Sigma$. Let λ_0' be a prime divisor of λ_0 in K. Suppose that λ_0' splits in K_λ, i.e. $\mathrm{Fr}_{\lambda_0'} = id$.

The extension $K_{\lambda_0\lambda}/K_\lambda$ is a cyclic extension of degree $\lambda_0 - \left(\frac{\lambda_0}{K}\right)$. Recall that

$$m_{\lambda_0} := \mathrm{ord}_\ell\left(\gcd\left(\lambda_0 - \left(\frac{\lambda_0}{K}\right), a_{\lambda_0} - 1 - \left(\frac{\lambda_0}{K}\right)\right)\right).$$

Put $M_{\lambda_0} = \ell^{m_{\lambda_0}}$. Let's abbreviate M_{λ_0} as M in this section. Since $M \mid \lambda_0 - \left(\frac{\lambda_0}{K}\right)$, there exists an extension L of K_λ inside $K_{\lambda_0\lambda}$ of degree M. Let t_{λ_0} be a generator of the cyclic group $G(L/K_\lambda)$ which will be abbreviated as t, that is, $G(L/K_\lambda) = \langle t \rangle$.

Let M_1 be any power of ℓ, if $k = 0$. For $k > 0$, let $M_1 = M_{\lambda_1} = \ell^{m_{\lambda_1}}$. Suppose that

$$M | M_1 \quad \text{and} \quad \frac{M_1}{M} E(K_\lambda)_{M_1} \subset (t-1)E(L)_M. \tag{10}$$

Definition 13 For $\tau_{\lambda_0\lambda,\pi}(M_1), \tau_{\lambda,\pi}(M_1) \in (T, \Sigma)$, a k-th order Euler system, we define

$$\tau_1 := \mathrm{Cor}_{K_{\lambda_0\lambda}/L}\left(\tau_{\lambda_0\lambda,\pi}(M_1)\right) - \frac{\mathcal{L}_{\lambda_0}}{M} \cdot \tau_{\lambda,\pi}(M_1) \in H^1(L, E_{M_1}).$$

Remark 14 Notice the similarity between τ_1 and P_L in Definition 5. Indeed, if we compute $\mathrm{Cor}_{L/K_\lambda} \tau_1 \in H^1(K_\lambda, E_{M_1})$, we get

$$\begin{aligned}\mathrm{Cor}_{L/K_\lambda} \tau_1 &= \mathrm{Cor}_{K_{\lambda_0\lambda}/K_\lambda}\left(\tau_{\lambda_0\lambda,\pi}(M_1)\right) - \frac{\mathcal{L}_{\lambda_0}}{M} \cdot \mathrm{Cor}_{L/K_\lambda}\left(\tau_{\lambda,\pi}(M_1)\right) \\ &= \mathcal{L}_{\lambda_0}\tau_{\lambda,\pi}(M_1) - \frac{\mathcal{L}_{\lambda_0}}{M} \cdot [L : K_\lambda] \cdot \tau_{\lambda,\pi}(M_1) = 0.\end{aligned}$$

Let $\varphi : G(\overline{L}/L) \to E_{M_1}$ be a cocycle representing τ_1 and let $\varphi_1 : G(\overline{K}_\lambda/K_\lambda) \to E_{M_1}$ be a cocycle representing $\mathrm{Cor}_{L/K_\lambda} \tau_1$.

Fix $g_0 \in G(\overline{K}_\lambda/K_\lambda)$ such that $g_0|_L = t|_L$.

Since φ_1 is a coboundary, there exists $e \in E_{M_1}$ such that

$$\varphi_1(g_0) = (g_0 - 1)e \quad \text{and} \quad \varphi_1(h) = (h-1)e, \quad \forall h \in G(\overline{L}/L).$$

Let

$$e' = \frac{M_1}{M} e \in E_M \quad \text{and} \quad \varphi' = \frac{M_1}{M}\varphi.$$

Define a map $\varphi_2 : G(\overline{K}_\lambda/K_\lambda) \to E_M$ as follows:

$$\begin{cases} \varphi_2(id) = 0 \\ \varphi_2(g_0^i) = (g_0^i + \cdots + 1)e' \quad \text{for } 1 \geqslant i \geqslant M-1 \\ \varphi_2(h) = (M-1)\varphi'(h) + (M-2)g_0\varphi'(g_0^{-1}hg_0) + \cdots + g_0^{M-2}\varphi'(g_0^{-(M-2)}hg_0^{M-2}) \\ \varphi_2(hg_0^i) = h\varphi_2(g_0^i) + \varphi_2(h) \end{cases}$$

A somewhat time-consuming but elementary verification shows that φ_2 is a cocycle whose cohomology class does not depend on the choice of φ, e or g_0.

Definition 15 $\tau_{\lambda,\lambda_0,\pi}(M) \in H^1(K_\lambda, E_M)$ is the class of the cocycle φ_2 constructed above.

Definition 16 Suppose that $\lambda_0 \in \mathcal{P}(2N\lambda)$, $\alpha = [\lambda, \pi] \in \Sigma$, and λ_0' is a prime divisor of λ_0 in K. If $k = 0$, we say that the pair (α, λ_0) is *admissible* if $\lambda_0\lambda \in \Sigma$ and λ_0' splits in K_λ. If $k > 0$, we say that (α, λ_0) is *admissible* if $[\lambda_0\lambda, \pi] \in \Sigma$, λ_0' splits in K_λ, and satisfies (10).

Definition 17 $\Sigma' := \{[\lambda, \lambda_0, \pi] \mid ([\lambda, \pi], \lambda_0)$ runs through all admissible pairs $\}$.

Definition 18 $T' := \{\tau_\beta \mid \beta \in \Sigma'\}$.

Theorem 19 *T' is an Euler system.*

Proof We need to show : let $p \in \mathcal{P}(2N\lambda)$ and $[\lambda, \lambda_0, \pi] \in \Sigma'$ such that the prime divisor $\wp$ of p in K is unramified in K_λ and $[p\lambda, \lambda_0, \pi] \in \Sigma'$. Then

$$\mathrm{Cor}_{K_{p\lambda_0\lambda}/K_{\lambda_0\lambda}}(\tau_{p\lambda,\lambda_0,\pi}) = \mathcal{L}_p \tau_{\lambda,\lambda_0,\pi}.$$

5 Structure of Ш(K, E)

In this section, assume that E is an elliptic curve without complex multiplication and with square-free conductor N. For each rational prime p, fix a generator $t_p \in G(K_p/K_1)$ and let t_p also denote the generator of $G(K_\lambda/K_{\lambda/p})$ corresponding to it. Both of these Galois groups are cyclic of order $p+1$ if we assume that p is inert in K. For a rational prime $\ell \geqslant 11$, let $M = \ell^n$ for any $n \geqslant 1$ that makes M satisfy (7). Let

$$D_p = \sum_{j=1}^{p} j t_p^j \quad \text{and} \quad D_\lambda = \prod_{p|\lambda} D_p \in \mathbb{Z}[G(K_\lambda/K_1)].$$

Let S be a set of coset representatives of $G(K_\lambda/K)$ with respect to $G(K_\lambda/K_1)$. Let

$$P_\lambda = \sum_{\sigma \in S} \sigma(D_\lambda \mathcal{Y}_\lambda) \in E(K_\lambda).$$

The class $P_\lambda \pmod{ME(K_\lambda)}$ is independent of the choice of S and depends on the generator $t_p \in G(K_p/K_1)$ for $p|\lambda$.

Also, $P_\lambda \pmod{ME(K_\lambda)} \in E(K_\lambda)/ME(K_\lambda)$ can be considered as an element of $H^1(K_\lambda, E_M)$ through the Kummer exact sequence

$$O \to E(K_\lambda)/ME(K_\lambda) \to H^1(K_\lambda, E_M) \to H^1(K_\lambda, E)_M \to O.$$

It can be shown that $P_\lambda \pmod{ME(K_\lambda)}$ is fixed by $G(K_\lambda/K)$. The proof uses an induction on the number of prime divisors of λ, together with the norm relation. Therefore,

$$P_\lambda \pmod{ME(K_\lambda)} \in H^1(K_\lambda, E_M)^{G(K_\lambda/K)}.$$

The Hochschild-Serre spectral sequence gives us an isomorphism

$$\mathrm{Res}_{K_\lambda/K} : H^1(K, E_M) \xrightarrow{\sim} H^1(K_\lambda, E_M)^{G(K_\lambda/K)}.$$

Let $\tau_{\lambda,n}$ be the unique class in $H^1(K, E_M)$ such that

$$\mathrm{Res}_{K_\lambda/K}(\tau_{\lambda,n}) = P_\lambda \pmod{ME(K_\lambda)}.$$

The classes $\tau_{\lambda,n}$ can generate elements in $\text{Ш}(K, E)_{\ell^\infty}$, the ℓ-component of $\text{Ш}(K, E)$. On the other hand, the orthogonality relation between the elements of $H^1(K, E_M)$ and $\tau_{\lambda,n}$ restricts the size of $\text{Ш}(K, E)_{\ell^\infty}$.

References

1. J. Cassels, A. Fröhlich (eds.), *Algebraic Number Theory* (Academic Press, Cambridge, 1967)
2. B. Gross, Kolyvagin's work for modular elliptic curves, in *L-Functions and Arithmetic*, ed. by J. Coates, M.J. Taylor. London Mathematical Society Lecture Note Series, vol. 153 (Cambridge University Press, Cambridge, 1991), pp. 235–256
3. B. Gross, D. Zagier, Heegner points and derivatives of L-series. Invent. Math. **84**(2), 225–320 (1986)
4. V. Kolyvagin, Finiteness of $E(\mathbb{Q})$ and $\text{Ш}(E, \mathbb{Q})$ for a subclass of Weil curves. Izv. Akad. Nauk SSSR. Ser. Mat. **52**, 522–540 (1988) [English translation in Math. USSR Izv. **32**, 523–542 (1989)]
5. V. Kolyvagin, Mordell-Weil and Shafarevich-Tate groups. Izv. Akad. Nauk SSSR Ser. Mat. **52**, 1154–1180 (1988) [English translation in Math. USSR Izv. **33**, 474–499 (1989)]
6. V. Kolyvagin, Euler systems, in *The Grothendieck Festschrift*, vol. 2, ed. by P. Cartier, L. Illusie, N. Katz, G. Laumon, Y. Manin, K. Ribet. Progress in Mathematics, vol. 87 (Birkhäuser, Boston, 1990), pp. 435–483
7. V. Kolyvagin, On the structure of Shafarevich-Tate groups, *Algebraic Geometry*. Lecture Notes in Mathematics, vol. 1479 (Springer, Berlin, 1991), pp. 94–121
8. J. Silverman, *The Arithmetic of Elliptic Curves*, 2nd edn. (Springer, Berlin, 2009)

p-Adic Analogues of the BSD Conjecture and the $\mathcal{L}$-Invariant

Chandrakant Aribam and Narasimha Kumar

Abstract In this lecture notes, we give an introduction to the p-adic analogues of the Birch and Swinnerton-Dyer conjecture for elliptic curves over $\mathbb{Q}$, when p is a prime of split multiplicative reduction for the elliptic curve. We quickly go through the p-adic methods and the tools from Hida theory, state the exceptional zero conjecture, and give a sketch of the proof of a conjecture of Mazur, Tate and Teitelbaum on the first derivative of p-adic L-functions due to Greenberg and Stevens.

1 Introduction

Let E be an elliptic curve defined over $\mathbb{Q}$. It is well-known that corresponding to E, there exists a modular cusp form f_E of weight 2, which is an eigenform for all the Hecke operators [3, 21]. As a result, the Hasse-Weil L-function, $L(E, s)$ for $s \in \mathbb{C}$, is known to have analytic continuation at $s = 1$. Further, if $\widetilde{E}$ denotes the reduced elliptic curve over $\mathbb{F}_p$, then the p-th Fourier coefficient of f_E equals $a_p := 1 + p - |\widetilde{E}(\mathbb{F}_p)|$.

Let $L^{(k)}(E, 1) := (1/k!).\frac{d^k}{ds^k}L(E, s)\mid_{s=1}$ be the k-th derivative of $L(E, s)$ evaluated at $s = 1$. Then the following statement is the well-known Birch and Swinnerton-Dyer conjecture for elliptic curves E defined over $\mathbb{Q}$.

Conjecture 1.1 (Classical BSD) *Let $r := rank_{\mathbb{Z}} E(\mathbb{Q})$. Then*

$$ord_{s=1} L(E, s) = r \tag{1}$$

C. Aribam (✉)
Department of Mathematical Sciences, Indian Institute of Science Education and Research, Mohali Sector 81, Knowledge City 140306, Punjab, India
e-mail: aribam@iisermohali.ac.in

N. Kumar
Department of Mathematics, Indian Institute of Technology Hyderabad, Kandi, Sangareddy 502285, Telangana, India
e-mail: narasimha.kumar@iith.ac.in

Z. Liang and C. Aribam (eds.), *The Computational and Theoretical Aspects of Elliptic Curves*, Mathematical Lectures from Peking University,
https://doi.org/10.1007/978-981-13-6664-2_3

and

$$L^{(r+1)}(E,1) = |\text{Ш}_{E/\mathbb{Q}}|.\frac{R_\infty(E)}{|E(\mathbb{Q})_{\text{tor}}|^2}\left(\prod_\ell c_\ell\right)\Omega_E, \tag{2}$$

where $R_\infty(E)$ is the classical regulator of $E/\mathbb{Q}$, $\text{Ш}_{E/\mathbb{Q}}$ is the Tate-Shafarevich group of E over $\mathbb{Q}$, c_ℓ are the local Tamagawa factors, Ω_E is the positive real period of E, and $E(\mathbb{Q})_{\text{tor}}$ is the torsion group of $E(\mathbb{Q})$. By the well-known theorem of Mordell, the torsion group $E(\mathbb{Q})_{\text{tor}}$ is known to be finite.

Consider the p-adic L-function, $L_p(E,s) \in \mathbb{Q}_p[[s]]$, which interpolates the complex values of the L-function of E (see §2 below). Then we have the following conjecture.

Conjecture 1.2 (Mazur, Tate, Teitelbaum) *Let $ord_{s=1}L_p(E,s)$ be the order of vanishing of $L_p(E,s)$ at $s=1$. Then*

$$\text{ord}_{s=1}L_p(E,s) = \begin{cases} rank_{\mathbb{Z}}E(\mathbb{Q}) & \text{if } p \text{ is a prime of good ordinary reduction,} \\ rank_{\mathbb{Z}}E(\mathbb{Q})+1 & \text{if } p \text{ is a prime of split multiplicative reduction.} \end{cases} \tag{3}$$

Conjecture 1.3 (Mazur, Tate, Teitelbaum) *Let E be an elliptic curve defined over $\mathbb{Q}$ and p be a prime of split multiplicative reduction for E. Then $ord_{s=1}L_p(E,s) = rank_{\mathbb{Z}}E(\mathbb{Q})+1$ and the leading coefficient of $L_p(E,s)$ is given by*

$$L_p^*(E,s) = \mathcal{L}_p(E)|\text{Ш}_{E/\mathbb{Q}}|.\frac{\text{Reg}_p^{sch}(E)}{|E(\mathbb{Q})_{\text{tor}}|^2}\left(\prod_\ell c_\ell\right)\Omega_E, \tag{4}$$

where $\mathcal{L}_p(E)$ is the $\mathcal{L}$-invariant of $E/\mathbb{Q}_p$, $\text{Reg}_p^{sch}(E)$ is the p-adic regulator of E computed with respect to the Schneider height. The terms c_ℓ are the local Tamagawa factors of E, and $\text{Ш}_{E/\mathbb{Q}}$ is the Tate-Shafarevich group of E over $\mathbb{Q}$. All the terms involved are defined and beautifully explained in [19]. A conjecture for the cases when p is a prime of good ordinary or non-split multiplicative case is also given in [19, p. 38].

This conjecture is referred to as the p-adic analogue of the conjecture of Birch and Swinnerton-Dyer for elliptic curves over $\mathbb{Q}$.

Hida theory for Galois representations arising from congruences between modular forms and p-adic methods for elliptic curves have turned out to be very important tools for proving results about special values of p-adic L-functions of elliptic curves. Apart from explaining the p-adic analogues of BSD, one of the goals of the workshop was to demonstrate these tools and p-adic methods. For this, we focused on an interesting aspect of Conjecture 1.2, namely, the second part where one can see the occurrence of an extra zero of the p-adic L-function. The presence of this extra zero is attributed to the interpolation formula of the p-adic L-function. This phenomenon was first observed by Mazur, Tate and Teitelbaum, and they conjectured that the special value of the p-adic L-function at the central point is then related to the special value of the complex L-function through an invariant defined locally in terms of the Tate module of E over $\mathbb{Q}_p$. We recall the definition of this invariant below and its conjectured description due to Mazur, Tate and Teitelbaum, and a brief sketch of the proof due to Greenberg and Stevens.

Definition 1.4 Suppose the elliptic curve E have split multiplicative reduction at the prime p. Then the p-th Fourier coefficient of f_E is equal to 1. By Tate's uniformization theorem, we have an isomorphism

$$E(\overline{\mathbb{Q}}_p) \cong \overline{\mathbb{Q}}_p^{\times}/q_E^{\mathbb{Z}},$$

where $q_E \in \mathbb{Q}_p^{\times}$ is a p-adic Tate period. The $\mathcal{L}$-invariant $\mathcal{L}(E)$ is defined by

$$\mathcal{L}(E) = \frac{\log_p(q_E)}{\mathrm{ord}_p(q_E)}. \tag{5}$$

Then the conjecture of Mazur, Tate and Teitelbaum [19], proved by Greenberg and Stevens [6, Theorem 4.1], is as follows:

Theorem 1.5 *Let $p \geq 5$ be a prime and let E be an elliptic curve defined over $\mathbb{Q}$ with split multiplicative reduction at p. Let $L_p(E, s) \in \mathbb{Z}_p[[\Gamma]]$, with $s \in \Gamma \cong \mathbb{Z}_p$, be the p-adic L-function constructed by Mazur and Swinnerton-Dyer. Then*

$$\frac{dL_p(E, s)}{ds}\Big|_{s=1} = \mathcal{L}(E)\frac{L(E, 1)}{\Omega_E}, \tag{6}$$

where $L(E, z)$ is the Hasse-Weil L-function of E and Ω_E is the positive Neron period of E.

There are several proofs of the above theorem with different flavors. The first proof was given by Greenberg–Stevens [6] using a global theory like Hida's universal ordinary deformation, which we sketch a proof below. The other is, as Kato–Kurihara–Tsuji [14] or Colmez [4] did, a proof based on local theory (except using Kato's element). The p-adic L-function is the image of Kato's element via a purely local morphism, the so called Coleman map or Perrin-Riou map. The extra zero phenomena discovered by Mazur–Tate–Teitelbaum is, in fact, a property of the local Coleman map. An elementary proof of MTT conjecture for elliptic curves by using Kato's element is given by Kobayashi [15].

In [1], the author provides two formulas for Greenberg's generalized $\mathcal{L}$-invariant in terms of derivatives of eigenvalues of Frobenius. As a special case, the author obtains a new proof of the MTT Conjecture, but by using the cohomology of (φ, Γ)-modules instead of Galois cohomology. In the same article, a generalized conjecture on trivial zeros, which is referred as Trivial zeros conjecture, made for any n-dimensional continuous irreducible p-adic representation of $G_{\mathbb{Q}}$. In [2], the author uses the theory of Robba ring (φ, Γ)-modules to generalize Greenberg's construction of the $\mathcal{L}$-invariant to p-adic representations which are semi-stable at p. This conjecture is known to be true in the case of symmetric square representation of a modular form such that the associated automorphic representation is the Steinberg one (cf. [17, 20]).

The plan of the lectures is as follows. In the first lecture, we shall recall the p-adic L-function of an elliptic curve. We then say about the universal deformation of the Galois representation attached to an elliptic curve and we give the definition of the algebraic $\mathcal{L}$-invariant. In the second lecture, we give a formula for the $\mathcal{L}$-invariant in terms of the Fourier coefficients of a Λ-adic form. We use this formula to give a sketch of a proof of Theorem 1.5. For our presentation, we follow the exposition given at the end of Chap. 1 of Hida's book [12].

We have benefited greatly by the article of Greenberg and Stevens [6] as well as the articles by Hida on the subject [11–13].

2 p-Adic L-Function

Let $p \geq 5$ be a prime and $N \geq 1$ be an integer prime to p. We fix an embedding $\overline{\mathbb{Q}} \hookrightarrow \overline{\mathbb{Q}}_p$ throughout. As mentioned above, it is well-known that corresponding to E over $\mathbb{Q}$, there exists a modular cusp form f_E of weight 2, which is an eigenform for all the Hecke operators such that the L-function of f_E is equal to the L-function $L(E, s)$ [3, 21].

It was already known due to Eichler, Shimura, Deligne, and Deligne-Serre, that associated to any newform $f = \sum_{n \geq 1} a_n(k) q^n$ of weight $k \geq 1$, there exists a Galois representation $\rho_f : G_{\mathbb{Q}} \to \mathrm{GL}_2(\overline{\mathbb{Q}}_p)$, which is unramified away from the level of f and p. Then, by Wiles's theorem *loc. cit*, the Galois representation ρ_{f_E} and the Galois representation arising from the Tate module of the elliptic curve E are also naturally equivalent.

If f is ordinary at p, then the p-th Fourier coefficient $a_p(k)$ is a p-adic unit. It then follows that the polynomial $X^2 - a_p(k)X + p^{k-1}$ has a unique p-adic unit root which we denote by $\alpha_p(k)$. Further, the restriction of ρ_f to the decomposition subgroup G_p, one has

$$\rho_f\,|_{G_p} \sim \begin{pmatrix} \epsilon & * \\ 0 & \delta \end{pmatrix}$$

where ϵ, δ are characters with δ unramified.

Let p be a prime of ordinary reduction of E. Then the Galois representation associated to the p-adic Tate module of E is ordinary at p. Let $\mathbb{Q}_\infty$ denote the cyclotomic $\mathbb{Z}_p$-extension of $\mathbb{Q}$, and $\Gamma = \mathrm{Gal}(\mathbb{Q}_\infty/\mathbb{Q}) \cong \mathbb{Z}_p$. We fix a topological generator γ for Γ. We set $\Lambda := \mathbb{Z}_p[[T]]$, the power series in one variable, then Λ can be identified with the $\mathbb{Z}_p[[\Gamma]]$. Then a p-adic L-function has been constructed by Mazur and Swinnerton-Dyer such that it interpolates the special values of the Hasse-Weil L-function at 1 [18]. More precisely, there exists a function $\mathscr{L}_p(E, T) \in \Lambda \otimes_{\mathbb{Z}_p} \mathbb{Q}_p$ such that for every *non-trivial* character χ with $\chi(\gamma) = \zeta \in \mu_{p^\infty}$ and ζ of order p^m, we have the following interpolation property:

$$\mathscr{L}_p(E, \zeta - 1) = \tau(\chi^{-1})\alpha_p^{-m} . \frac{L(E, \chi, 1)}{\Omega_E}, \tag{7}$$

where $\tau(\chi^{-1})$ denotes the Gauss sum, $a_p := 1 + p - \#\tilde{E}(\mathbb{F}_p)$, and α_p is the unique p-adic unit root of the polynomial $X^2 - a_p X + p$. The function $L(E, \chi, s)$ is the Hasse-Weil L-function twisted by χ. By the Weierstrass preparation theorem, the power series $\mathscr{L}_p(E, T)$ is uniquely determined by the interpolation property (7).

Let $\chi_p : G_{\mathbb{Q}} \to \mathbb{Q}_p^\times$ denote the cyclotomic character defined by the action of $\mathrm{Gal}(\mathbb{Q}(\mu_{p^\infty})/\mathbb{Q})$ on μ_{p^∞}. Then $\chi_p \mid_\Gamma$ induces an isomorphism of Γ with $1 + p\mathbb{Z}_p$. Then the p-adic L-function of E at the prime p is defined by

$$L_p(E, s) = \mathscr{L}_p(E, \chi_p^{1-s}(\gamma) - 1).$$

Further, $L_p(E, 1) = \mathscr{L}_p(E, 0)$. If χ is the trivial character and p is a prime of good reduction for E, then

$$L_p(E, 1) = \mathscr{L}_p(E, 0) = (1 - \alpha_p^{-1})^2 \frac{L(E, 1)}{\Omega_E},$$

where α_p is as above.

The p-adic L-function $L_p(E, s)$ is in $\mathbb{Q}_p[[s]]$ and is independent of the choice of the topological generator γ. In fact, in certain cases it can be shown that it is in $\mathbb{Z}_p[[s]]$. For instance, if $E[p]$ is absolutely irreducible as a representation of $G_{\mathbb{Q}}$, then by [9, Prop 3.7], the p-adic L-function $L_p(E, s) \in \mathbb{Z}_p[[s]]$.

Let ϵ_∞ denote the sign in the functional equation for the Hasse-Weil L-function $L(E, z)$. If E has split multiplicative reduction at p, then the p-adic L-function $L_p(E, s)$ has the functional equation

$$L_p(E, 2 - s) = \epsilon_p \langle N \rangle^{s-1} L_p(E, s).$$

Here, $\epsilon_p = -\epsilon_\infty$, N is the conductor of E and for the Teichmüller character ω, $\langle N \rangle = N\omega^{-1}(N)$. If $\epsilon_\infty = -1$, then $\epsilon_p = +1$ and the quantities on both sides of (6) are zero by the above formula for $L_p(E, 1)$. Therefore, Theorem 1.5 holds trivially. To prove the theorem in the remaining case, we need some machinery, in particular, a two variable p-adic L-function associated to E.

In [16], Kitagawa constructed a two-variable p-adic L-function $L_p(k, s)$ associated to E, which is analytic in a neighborhood of 2 in $\mathbb{Z}_p$. The first variable is the "weight" variable and the second is the "cyclotomic" variable. This two-variable p-adic L-function has the following properties:

(i) $L_p(2, s) = L_p(E, s)$ for all $s \in \mathbb{Z}_p$;
(ii) $L_p(k, k - s) = \epsilon_p \langle N \rangle^{s-k/2} L_p(k, s)$, for all $s \in \mathbb{Z}_p$;

(iii) $L_p(k,1) = (1 - a_p(k)^{-1})L_p^*(k)$ where $L_p^*(k)$ is a p-adic analytic function of k such that $L_p^*(2) = \dfrac{L(E,1)}{\Omega_E}$, and $a_p(k)$ is the p-th Fourier coefficient of a weight k cusp form;

This two-variable p-adic L-function plays a key role in the proof of Theorem 1.5.

3 Hida Theory

Let $p \geq 5$ be a prime of ordinary reduction for E which is defined $\mathbb{Q}$. Let $\mathbb{T}$ be the ordinary Hecke algebra of tame level N constructed by Hida [10], and $\mathbb{I}$ be the normalization of an irreducible component of the fraction field of $\mathbb{T}$. It is a well-known theorem due to Hida, that there exists a representation of $G_{\mathbb{Q}}$ into $GL_2(\mathbb{I})$. By the process of specialization, this representation gives rise to the representations attached to p-adic cusp forms of every weight, and in particular the weight 2 specialization gives rise to ρ_E. Further all the Galois representations obtained through specialization are congruent modulo p to the representation ρ_E.

Throughout, we shall assume the following:

$$\mathbb{I} = \Lambda := \mathbb{Z}_p[[1 + p\mathbb{Z}_p]],$$

which is isomorphic to the Iwasawa algebra in one variable. Then the formal power series $\mathcal{F}(q) = \sum_{n=1}^{\infty} \mathbf{a}(n)q^n \in \mathbb{I}[[q]]$, where $\mathbf{a}(n) \in \mathbb{I}$ is the image of a Hecke operator $T(n)$ in $\mathbb{I}$, is an $\mathbb{I}$-adic form. By duality between Hecke algebras and the space of cusp forms followed by specialization, this $\mathbb{I}$-adic form gives rise to p-adic eigenforms for each weight. The $\mathbb{I}$-adic form $\mathcal{F}$ is also referred to as the $\mathbb{I}$-adic lift of the modular form f_E.

Consider the universal ordinary deformation ring $\mathcal{R}$ of the Galois representation ρ_{f_E}. It is a theorem due to Wiles and Taylor-Wiles, that there is a local isomorphism between $\mathcal{R}$ and $\mathbb{I}$. Hida theory is now available for the prime $p = 3$ by the work of Wiles and for $p = 2$ by Ghate-Kumar [5].

4 Definition of $\mathcal{L}$-Invariant

Let V denote the representation space of ρ_E which is defined over $\mathbb{Q}_p$. The adjoint representation $\mathrm{Ad}^0(V)$ consisting of trace zero matrices plays a crucial role in proving Theorem 1.5.

We begin by recalling the definition of an algebraic invariant called $\mathcal{L}^{alg}$, due to Greenberg in [8]. By the ordinarity of V at p, we have a short exact sequence of G_p-modules:

$$0 \longrightarrow V^+ \longrightarrow V \longrightarrow V^- \longrightarrow 0.$$

Identifying the elements of $\mathrm{End}(V)$ with matrices with respect to a basis of V containing a generator of V^+, we consider the following subspaces:

(i) $\mathrm{Ad}^0(V)^-$ consisting of upper triangular matrices with trace zero,
(ii) $\mathrm{Ad}^0(V)^+$ consisting of upper nilpotent matrices of trace zero.

Then it is easy to see that the decomposition group G_p acts trivially on the quotient $\mathrm{Ad}^0(V)^-/\mathrm{Ad}^0(V)^+$, i.e., $\mathrm{Ad}^0(V)^-/\mathrm{Ad}^0(V)^+ \cong \mathbb{Q}_p$ as G_p-modules.

Consider the maps

$$H^1(\mathbb{Q}_p, \mathrm{Ad}^0(V)^+) \longrightarrow H^1(\mathbb{Q}_p, \mathrm{Ad}^0(V)) \longrightarrow H^1(I_p, \mathrm{Ad}^0(V)/\mathrm{Ad}^0(V)^+),$$

and set

(i) $H^1(\mathbb{Q}_p, \mathrm{Ad}^0(V))^+ :=$ the image of $H^1(\mathbb{Q}_p, \mathrm{Ad}^0(V)^+) \longrightarrow H^1(\mathbb{Q}_p, \mathrm{Ad}^0(V))$
(ii) $U_p(\mathrm{Ad}^0(V)) := \ker\left(H^1(\mathbb{Q}_p, \mathrm{Ad}^0(V)) \longrightarrow H^1(I_p, \mathrm{Ad}^0(V)/\mathrm{Ad}^0(V)^+)\right)$.

Then $H^1(\mathbb{Q}_p, \mathrm{Ad}^0(V))^+ \subset U_p(\mathrm{Ad}^0(V))$. Similarly, we set

(i) $H^1(\mathbb{Q}_p, \mathrm{Ad}^0(V))^- :=$ the image of $H^1(\mathbb{Q}_p, \mathrm{Ad}^0(V)^-) \longrightarrow H^1(\mathbb{Q}_p, \mathrm{Ad}^0(V))$.

Then using these cohomology groups over local fields, Greenberg defined the following Selmer group, which is referred to as a *balanced Selmer group* in [12]:

$$\mathrm{Sel}_{\mathbb{Q}}(\mathrm{Ad}^0(V)) := \ker\left[H^1(\mathbb{Q}, \mathrm{Ad}^0(V)) \longrightarrow \prod_{q \neq p} H^1(\mathbb{Q}_q, \mathrm{Ad}^0(V)) \oplus \frac{H^1(\mathbb{Q}_p, \mathrm{Ad}^0(V))}{H^1(\mathbb{Q}_p, \mathrm{Ad}^0(V)^+)}\right]. \tag{8}$$

Proposition 4.1 ([6])

$$\mathrm{Sel}_{\mathbb{Q}}(\mathrm{Ad}^0(V)) = 0. \tag{9}$$

As a consequence, we have

$$H^1(\mathbb{Q}, \mathrm{Ad}^0(V)) \cong \frac{H^1(\mathbb{Q}_p, \mathrm{Ad}^0(V))}{H^1(\mathbb{Q}_p, \mathrm{Ad}^0(V))^+}.$$

Therefore, there exists a unique subspace $\mathbf{H}$ of $H^1(\mathbb{Q}_p, \mathrm{Ad}^0(V))$ projecting onto

$$\frac{H^1(\mathbb{Q}_p, \mathrm{Ad}^0(V))^-}{H^1(\mathbb{Q}_p, \mathrm{Ad}^0(V))^+} \hookrightarrow \frac{H^1(\mathbb{Q}_p, \mathrm{Ad}^0(V))}{H^1(\mathbb{Q}_p, \mathrm{Ad}^0(V))^+} \cong H^1(\mathbb{Q}, \mathrm{Ad}^0(V)).$$

Then by restriction, we have

$$\mathbf{H} \hookrightarrow \mathrm{Hom}(G_p^{ab}, \mathrm{Ad}^0(V)^-/\mathrm{Ad}^0(V)^+).$$

As $\mathrm{Ad}^0(V)^-/\mathrm{Ad}^0(V)^+ \cong \mathbb{Q}_p$ as G_p-modules, with a trivial action, we have

$$\mathrm{Hom}(G_p^{ab}, \mathrm{Ad}^0(V)^-/\mathrm{Ad}^0(V)^+) \cong \mathrm{Hom}(\mathrm{Gal}(M_\infty^+/\mathbb{Q}_p), \mathbb{Q}_p) \times \mathrm{Hom}(\mathrm{Gal}(M_\infty^{ur}/\mathbb{Q}_p), \mathbb{Q}_p) \cong \mathbb{Q}_p^2,$$

where M_∞^+ is the cyclotomic $\mathbb{Z}_p$-extension of $\mathbb{Q}_p$, and M_∞^{ur} is the unique unramified extension of $\mathbb{Q}_p$.

The one dimensional spaces $\mathrm{Hom}(\mathrm{Gal}(M_\infty^+/\mathbb{Q}_p), \mathbb{Q}_p)$ and $\mathrm{Hom}(\mathrm{Gal}(M_\infty^{ur}/\mathbb{Q}_p), \mathbb{Q}_p)$ each have a natural basis given by $\log_p \circ \chi_p$ and ord_p (sending the Frobenius $Frob_p$ to 1) respectively. The last isomorphism is given by the canonical map $\phi \mapsto \left(\frac{\phi([u, \mathbb{Q}_p])}{\log_p \circ \chi_p([u, \mathbb{Q}_p])}, \phi([p, \mathbb{Q}_p]) \right)$ for any $u \in \mathbb{Z}_p^\times$ of infinite order, and $[u, \mathbb{Q}_p]$ is the local Artin symbol. Here $\chi_p([u, \mathbb{Q}_p]) = u^{-1}$, so the above map can be written as $\phi \mapsto \left(\frac{\phi([u, \mathbb{Q}_p])}{-\log_p(u)}, \phi([p, \mathbb{Q}_p]) \right)$. Note that the negative sign does not appear in [7, p. 421], as the notation $Frob_p$ is used for the inverse of the Artin symbol at p. We believe that the isomorphism in [12, Sect. 1.5.2, p. 64] should also have a negative sign in the first coordinate. We would like to thank the referee for pointing this out.

By Proposition 4.1, the projection to the first coordinate via $\phi \mapsto -\frac{\phi([u, \mathbb{Q}_p])}{\log_p(u)}$ is surjective. Then, it follows that the 1-dimensional image of $\mathbf{H}$ is a graph of a $\mathbb{Q}_p$-linear map

$$\mathcal{L}^{alg} : \mathrm{Ad}^0(V)^-/\mathrm{Ad}^0(V)^+ \longrightarrow \mathrm{Ad}^0(V)^-/\mathrm{Ad}^0(V)^+,$$

which is given by multiplication by an element

$$\mathcal{L}^{alg}(\mathrm{Ad}^0(V)) = \frac{\phi([p, \mathbb{Q}_p])}{\frac{\phi([u, \mathbb{Q}_p])}{-\log_p(u)}} \in \mathbb{Q}_p. \tag{10}$$

Assume now that p is a prime of split multiplicative reduction. Let q_E be the Tate period of $E(\overline{\mathbb{Q}}_p)$. From above, we have

$$\text{image of } \mathbf{H} \subset \mathrm{Hom}(G_p^{ab}, \mathrm{Ad}^0(V)^-/\mathrm{Ad}^0(V)^+) \cong \mathrm{Hom}(G_p^{ab}, \mathbb{Q}_p)$$

and the image is generated by a map $\phi_0 : G_p^{ab} \longrightarrow \mathbb{Q}_p$. It is easy to see that the field extension M_∞ of $\mathbb{Q}_p$ cut out by ϕ_0 is totally ramified and the *universal field of norms* has a rank 1 torsion free part, say q_0. Then the following result was shown by Greenberg in [8]. For our exposition, we follow the proof in [12, Proposition 1.85].

Proposition 4.2 *Let p be a prime of split multiplicative reduction for E and q_E be its Tate period. Then* $\mathcal{L}^{alg}(\mathrm{Ad}^0(V)) = \dfrac{\log_p(q_0)}{\mathrm{ord}_p(q_0)} = \dfrac{\log_p(q_E)}{\mathrm{ord}_p(q_E)}$.

Proof (*Sketch*): The equality is shown by showing that the two invariants are related to a third invariant denoted by $\mathcal{L}(V)$ (see [7, Def 3.9]).

For the first equality, consider the compositum $\mathbf{M}_\infty$ of all $\mathbb{Z}_p$-extensions of $\mathbb{Q}_p$. Then, by local class field theory, $\mathrm{Gal}(\mathbf{M}_\infty/\mathbb{Q}_p) \cong \mathbb{Z}_p^2$ and the Artin symbol $[q_0, \mathbb{Q}_p] \in \mathrm{Gal}(\mathbf{M}_\infty/M_\infty)$. Note that $\phi_0([q_0, \mathbb{Q}_p]) = 0$.

Let $q_0 = p^a u$, with $a = \mathrm{ord}_p(q_0)$. Then $\phi_0([q_0, \mathbb{Q}_p]) = a\phi_0([p, \mathbb{Q}_p]) + \phi_0([u, \mathbb{Q}_p]) = 0$. Let M_∞^{ur} be the unique unramified $\mathbb{Z}_p$-extension of $\mathbb{Q}_p$ and M_∞^{+} be the cyclotomic $\mathbb{Z}_p$-extension of $\mathbb{Q}_p$. Then $\phi_0 \mid_{\mathrm{Gal}(M_\infty^+/\mathbb{Q}_p)} = x \log_p \circ \chi_p$, for the cyclotomic character χ_p, and a constant $x \in \mathbb{Q}_p^\times$. As $\log_p(\chi_p([u, \mathbb{Q}_p])) = \log_p(u^{-1})$, we have

$$a\phi_0([p, \mathbb{Q}_p]) + x \log_p(u^{-1}) = 0.$$

It follows from the definition that

$$\mathcal{L}^{alg}(\mathrm{Ad}^0(V)) = -\frac{\phi_0([p, \mathbb{Q}_p])}{\phi_0([u, \mathbb{Q}_p])/\log_p(u)} = -\frac{\phi_0([p, \mathbb{Q}_p]) \log_p(u)}{\phi_0([u, \mathbb{Q}_p])} = \frac{\log_p(u)}{a} = \frac{\log_p(q_0)}{\mathrm{ord}_p(q_0)}.$$

By Tate's uniformization, we have $E(\overline{\mathbb{Q}}_p) \cong \overline{\mathbb{Q}}_p^\times/q_E^{\mathbb{Z}}$. This induces the following exact sequence of G_p-modules:

$$0 \longrightarrow \mu_{p^n} \longrightarrow E[p^n] \longrightarrow \mathbb{Z}/p^n\mathbb{Z} \longrightarrow 0.$$

From this sequence, by Cartier duality and taking the projective limit, we get the following sequence:

$$0 \longrightarrow \mathbb{Q}_p(1) \longrightarrow V_E \longrightarrow \mathbb{Q}_p \longrightarrow 0. \tag{11}$$

Again, splitting G_p^{ab} into two copies of $\mathbb{Z}_p$: into the maximal unramified $\mathbb{Z}_p$-extension and the cyclotomic $\mathbb{Z}_p$-extension, we have

$$H^1(\mathbb{Q}_p, \mathbb{Q}_p) = \mathrm{Hom}(G_p^{ab}, \mathbb{Q}_p) \cong \mathrm{Hom}(\mathrm{Gal}(M_\infty^+/\mathbb{Q}_p), \mathbb{Q}_p) \times \mathrm{Hom}(\mathrm{Gal}(M_\infty^{ur}/\mathbb{Q}_p), \mathbb{Q}_p).$$

The one dimensional spaces $\mathrm{Hom}(\mathrm{Gal}(M_\infty^+/\mathbb{Q}_p), \mathbb{Q}_p)$ and $\mathrm{Hom}(\mathrm{Gal}(M_\infty^{ur}/\mathbb{Q}_p), \mathbb{Q}_p)$ each have a natural basis given by $\log_p \circ \chi_p$ and ord_p (sending the Frobenius $Frob_p$ to 1) respectively. Therefore, they define isomorphisms:

$$\begin{aligned} i_+ &: \mathbb{Q}_p \longrightarrow \mathrm{Hom}(\mathrm{Gal}(M_\infty^+/\mathbb{Q}_p), \mathbb{Q}_p) : x \mapsto x.\log_p \circ \chi_p \\ i_{ur} &: \mathbb{Q}_p \longrightarrow \mathrm{Hom}(\mathrm{Gal}(M_\infty^{ur}/\mathbb{Q}_p), \mathbb{Q}_p) : x \mapsto x.\mathrm{ord}_p, \end{aligned} \tag{12}$$

gives the isomorphism: $H^1(\mathbb{Q}_p, \mathbb{Q}_p) \overset{(i_+^{-1}, i_{ur}^{-1})}{\cong} \mathbb{Q}_p^2$.

Consider the Tate duality of Galois cohomology

$$H^i(\mathbb{Q}_p, \mathbb{Q}_p) \times H^{2-i}(\mathbb{Q}_p, \mathbb{Q}_p(1)) \longrightarrow \mathbb{Q}_p. \tag{13}$$

For any $x \in \mathbb{Q}_p^\times$, and for any $n \in \mathbb{N}$, consider the cocyle given by $\sigma \mapsto (x^{1/p^n})^{\sigma-1}$ in $H^1(\mathbb{Q}_p, \mu_{p^n})$. Then the projective limit of these sequence of cocyles gives an element $\gamma_x \in H^1(\mathbb{Q}_p, \mathbb{Q}_p(1))$. Under the above pairing, the class γ_x gives a vector in the dual of $H^1(\mathbb{Q}_p, \mathbb{Q}_p)$. Let i_+^*, i_{ur}^* be the respective *dual* of the maps i_+, i_{ur} with respect to the above pairing. Then, identifying $\mathbb{Q}_p^*$ with $\mathbb{Q}_p$ by evaluation at 1, we get $i_+^*(\gamma_x)(1) = \log_p \circ \chi_p([x, \mathbb{Q}_p])$ and $i_{ur}^*(\gamma_x) = \mathrm{ord}_p(x)$.

Note that, the long exact sequence coming from the short exact sequence in (11) gives rise to the connecting morphism $\delta : H^1(\mathbb{Q}_p, \mathbb{Q}_p) \longrightarrow H^2(\mathbb{Q}_p, \mathbb{Q}_p(1)) \cong \mathbb{Q}_p$. Taking duals with respect to the above pairing, we get the morphism $\delta^* : H^0(\mathbb{Q}_p, \mathbb{Q}_p) \longrightarrow H^1(\mathbb{Q}_p, \mathbb{Q}_p(1))$. In fact, $\delta^*(y) = \gamma_y$. It follows that δ induces an isomorphism:

$$\mathrm{Hom}(\mathrm{Gal}(M_\infty^{ur}/\mathbb{Q}_p), \mathbb{Q}_p) \cong H^2(\mathbb{Q}_p, \mathbb{Q}_p(1)) = \mathbb{Q}_p. \tag{14}$$

Putting $\delta_{ur} = \delta \circ i_{ur}$ and $\delta_+ = \delta \circ i_+$, we have, by Greenberg's definition of the algebraic $\mathcal{L}$-invariant [7, Def 3.9]:

$$\mathcal{L}(V) = -\delta_{ur}^{-1} \circ \delta_+ = -i_{ur}^{-1} \circ i_+.$$

Taking the dual gives

$$\mathcal{L}(V) = -\frac{i_+^*(\gamma_{q_E})(1)}{i_{ur}^*(\gamma_{q_E})(1)} = \frac{\log_p(q_E)}{\mathrm{ord}_p(q_E)} = \mathcal{L}(E).$$

To complete the proof, we determine the image of the one dimensional space $\mathbf{H}$. Consider an inhomogeneous cocyle $c : G_{\mathbb{Q}} \longrightarrow \mathrm{Ad}^0(V)$, given by:

$$c(\sigma) = \begin{pmatrix} -a(\sigma) & b(\sigma) \\ 0 & a(\sigma) \end{pmatrix}, \forall \sigma \in G_p.$$

Now if the restriction to G_p modulo upper nilpotent cocycles is unramified, then $c(\sigma)$ gives rise to a non-trivial element of $\mathrm{Sel}_{\mathbb{Q}}(\mathrm{Ad}^0(V))$. Therefore, by Proposition 4.1, $a \mid_{I_p} \neq 0$. It follows that $\mathcal{L}^{alg}(\mathrm{Ad}^0(V)) = a([p, \mathbb{Q}_p]).\dfrac{\log_p(\gamma)}{a([\gamma, \mathbb{Q}_p])}$. Then, this is used to give the description of $\mathcal{L}^{alg}(\mathrm{Ad}^0(V))$ in Eq. (15) in the theorem below. Combining this with [6, Theorem 3.14], which also gives the description $\mathcal{L}(V) = a([p, \mathbb{Q}_p]).\dfrac{\log_p(\gamma)}{a([\gamma, \mathbb{Q}_p])}$, the result follows. □

We now relate $\mathcal{L}^{alg}(\mathrm{Ad}^0(V))$ to the derivative of the Fourier coefficient of an $\mathbb{I}$-adic form.

Theorem 4.3 *Let ρ_E denote the 2-dimensional Galois representation over $\mathbb{Q}_p$ of the elliptic curve E, and V denote the representation space. We assume that E has a split multiplicative reduction at a prime $p \geq 5$ and the residual representation is absolutely irreducible over $\mathbb{F}_p$. Let f_E denote the weight 2 Hecke eigenform associated to E. Let $\mathcal{F} = \sum_{n=1}^{\infty} \mathbf{a}(n)q^n \in \mathbb{I}[[q]]$ be the $\mathbb{I}$-adic lift of f_E. Suppose that the universal ordinary deformation ring $\mathcal{R}$ is isomorphic to the Hecke algebra $\mathbb{T}$. For $\Gamma = 1 + p\mathbb{Z}_p$, we fix an isomorphism $\mathbb{Z}_p[[\Gamma]] \cong \mathbb{Z}_p[[x]]$. Then*

$$\mathcal{L}^{alg}(\mathrm{Ad}^0(V)) = -2\log_p(\gamma)\frac{d\mathbf{a}(p)}{dx}\,|_{x=0}\,. \tag{15}$$

Proof (*Sketch*): Consider the Selmer group defined by

$$\mathrm{Sel}_{\mathbb{Q}}(\mathrm{Ad}^0(V))^{-} := \ker\left[H^1(\mathbb{Q}, \mathrm{Ad}^0(V)) \longrightarrow \prod_{q \neq p} H^1(\mathbb{Q}_q, \mathrm{Ad}^0(V)) \oplus \frac{H^1(\mathbb{Q}_p, \mathrm{Ad}^0(V))}{H^1(\mathbb{Q}_p, \mathrm{Ad}^0(V)^{-})}\right].$$

Recall that the subspace $\mathbf{H}$ is one-dimensional and is made up of classes unramified outside p and upper triangular on G_p. This shows that $\mathbf{H} \subset \mathrm{Sel}_{\mathbb{Q}}(\mathrm{Ad}^0(V))^{-}$ and by the vanishing of $\mathrm{Sel}_{\mathbb{Q}}(\mathrm{Ad}^0(V))$, we have $\mathrm{Sel}_{\mathbb{Q}}(\mathrm{Ad}^0(V))^{-} = \mathbf{H} \oplus \mathrm{Sel}_{\mathbb{Q}}(\mathrm{Ad}^0(V)) = \mathbf{H}$.

The good thing about the "minus" Selmer group is that we have the following isomorphism of the $\mathbb{Q}_p$-dual:

$$(\mathrm{Sel}_{\mathbb{Q}}(\mathrm{Ad}^0(V))^{-})^{*} \cong \Omega^1_{\mathcal{R}/\mathbb{Z}_p} \otimes_{\mathcal{R}} \mathcal{R}_P/P\mathcal{R}_P, \tag{16}$$

where $\mathcal{R}$ is the universal ordinary deformation ring. It is well-known that $\mathcal{R}$ is isomorphic to a Hecke algebra [21]. Let $\mathbb{I}$ be the irreducible component of the Hecke algebra to which ρ_E belongs, i.e., the ideal $P = (1 + x - (1 + p)^2)$ in $\mathbb{I}$ is the weight 2 arithmetic point corresponding to the representation ρ_E. Then the local ring $\mathcal{R}_P$ which is isomorphic to $\mathbb{Q}_p[[x]]$.

From the isomorphism (16), $\mathbf{H}$ is isomorphic to the tangent space at P of the localization $\mathcal{R}_P$, i.e., $\mathbf{H} \cong P\mathcal{R}_P/P^2\mathcal{R}_P$. In fact, the isomorphism is given by mapping an inhomogeneous cocyle c to an infinitesimal nearly ordinary deformation $\widetilde{\rho}_c : G_{\mathbb{Q}} \longrightarrow \mathrm{GL}_2(\mathbb{Q}_p[x]/(x^2))$, with determinant $\det(\widetilde{\rho}_c) = \det(\rho)$. As $\mathbf{H}$ is the tangent space at P, the element $(d\boldsymbol{\rho}/dx)\boldsymbol{\rho}^{-1}$ gives rise to a generator of $\mathbf{H}$, where $\boldsymbol{\rho}$ denotes the Galois representation associated to the universal deformation ring $\mathcal{R}_P$.

Now we take a non-trivial inhomogeneous cocycle $c \in \mathbf{H}$ as in the proof above. Let

$$c(\sigma) = \begin{pmatrix} -a(\sigma) & b(\sigma) \\ 0 & a(\sigma) \end{pmatrix}, \forall \sigma \in G_p.$$

Now if the restriction to the inertia group c modulo upper nilpotent cocyles is unramified, then $c(\sigma)$ gives rise to a non-trivial element of $\mathrm{Sel}_{\mathbb{Q}}(\mathrm{Ad}^0(V))$. Therefore $a\mid_{I_p}\neq 0$.

Therefore,

$$c(\sigma)\rho(\sigma) = C.\frac{d\rho}{dx}(\sigma),$$

where $C \in \mathbb{Q}_p^\times$ is a constant. For $\sigma \in G_p$, let $\rho(\sigma) = \begin{pmatrix} \epsilon(\sigma) & \beta(\sigma) \\ 0 & \delta(\sigma) \end{pmatrix}$. Then for the generator $\gamma = 1 + p$ of $1 + p\mathbb{Z}_p$, we have

$$a([p, \mathbb{Q}_p])\delta([p, \mathbb{Q}_p]) = C\frac{d\delta([p, \mathbb{Q}_p])}{dx}\mid_{x=0},$$

and

$$a([\gamma, \mathbb{Q}_p])\delta([\gamma, \mathbb{Q}_p]) = C\frac{d\delta([\gamma, \mathbb{Q}_p])}{dx}\mid_{x=0}.$$

If the universal deformation character of the trivial character is κ, then one can show that $\delta([\gamma, \mathbb{Q}_p]) = \kappa^{-1/2}$ ([12, p. 69]). Therefore

$$\frac{d\delta([\gamma^s, \mathbb{Q}_p])}{dx}\mid_{x=0} = -\frac{s}{2} \text{ and } \frac{d\delta([\gamma, \mathbb{Q}_p])}{dx}\mid_{x=0} = -\frac{1}{2}.$$

Combining with the fact that $\delta([p, \mathbb{Q}_p]) = \mathbf{a}(p)$, we get the result. □

5 The Proof of Theorem 1.5

We first explain the key ingredient in the proof of Theorem 1.5. Recall that $\mathbb{I} = \mathbb{Z}_p[[\Gamma]]$. First consider the lift of f_E to a unique $\mathbb{I}$-adic Hecke eigenform [10], say $\mathcal{F} = \sum_{n=1}^{\infty} \mathbf{a}(n)q^n \in \mathbb{I}[[q]]$. Then the key idea in the proof of the theorem is the following equality from [6, Theorem 3.18]:

$$\mathcal{L}(E) = -2\frac{d\mathbf{a}(p)}{dx}\mid_{x=0}, \tag{17}$$

where γ is the generator of the $\Gamma = 1 + p\mathbb{Z}_p$ under the isomorphism $\mathbb{Z}_p[[\Gamma]] \cong \mathbb{Z}_p[[x]]$.

If $\epsilon_p = 1$, then the p-adic L-function has an even order at $s = 1$ and the complex L-function $L(E, s)$ has an odd order at $s = 1$. Therefore, the theorem is trivially true in this case as both the sides vanish.

Now let $\epsilon_p = -1$. Let $L_p(k, s)$ be the two variable p-adic L-function. Note that, for $k \in \mathbb{Z}_p$, $L_p(k, k/2) = 0$ by the functional equation. This means that the linear terms in the Taylor expansion of $L_p(k, s)$ around the point $(k, s) = (2, 1)$ is zero along the line $s = k/2$. Hence, there is a constant $c \in \mathbb{Z}_p$ such that

$$L_p(k, s) = c\left((s-1) - \frac{k-2}{2}\right) + \text{ higher order terms in } s.$$

Here $c \neq 0$, as the weight 2 specialization of $L_p(k, s)$ is non-trivial. If $k = 2$, then $L_p(E, s) = c(s-1) +$ higher order terms. We then have,

$$c = \frac{dL_p(E, s)}{ds}\Big|_{s=1}. \tag{18}$$

On the other hand, if $s = 1$, then

$$(1 - \mathbf{a}_p(k)^{-1})L_p^*(k, 1) = -\frac{1}{2}c(k-2).$$

As p is a prime of split multiplicative reduction, we have $\mathbf{a}_p(2) = 1$. Differentiating the previous equality with respect to k and putting $k = 2$ yields

$$-\frac{1}{2}c = \mathbf{a}'_p(2)L_p^*(2, 1).$$

Now, by using the fact that $\mathbf{a}'_p(2) = -\frac{1}{2}\mathcal{L}(E)$ and $L_p^*(2, 1) = \frac{L(E, 1)}{\Omega_E}$, we have

$$-\frac{1}{2}c = -\frac{1}{2}\mathcal{L}(E).\frac{L(E, 1)}{\Omega_E}.$$

Combining this with (18), the theorem follows. □

Acknowledgements The authors would like to thank the organizers of the workshop and conference warmly, especially Liang Zhibin and Rouchuan Liu, for arranging the stay, and also BICMR for the hospitality and the local support. It was a nice little conference with lots of interaction among the participants. The idea to have this workshop and conference came up from the discussions between Sujatha and Liang, and it would not have happened without the support of Sujatha. We thank her warmly for her support and advice.

References

1. D. Benois, Infinitesimal deformations and the ℓ-invariant. Doc. Math. 5–31 (2010) (Extra Volume: Andrei A. Suslin sixtieth birthday)
2. D. Benois, A generalization of Greenberg's $\mathcal{L}$-invariant. Am. J. Math. **133**(6), 1573–1632 (2011)
3. C. Breuil, B. Conrad, F. Diamond, R. Taylor, On the modularity of elliptic curves over $\mathbb{Q}$: wild 3-adic exercises. J. Am. Math. Soc. **14**(4), 843–939 (2001)

4. P. Colmez, La conjecture de Birch et Swinnerton-Dyer p-adique. Astérisque **294**, ix, 251–319 (2004)
5. E. Ghate, N. Kumar, Control theorems for ordinary 2-adic families of modular forms, *Automorphic Representations and L-Functions*. Tata Institute of Fundamental Research Studies in Mathematics, vol. 22 (Tata Institute of Fundamental Research, Mumbai, 2013), pp. 231–261
6. R. Greenberg, G. Stevens, p-adic L-functions and p-adic periods of modular forms. Invent. Math. **111**(2), 407–447 (1993)
7. R. Greenberg, G. Stevens, On the conjecture of Mazur, Tate, and Teitelbaum, *p-Adic Monodromy and the Birch and Swinnerton-Dyer Conjecture (Boston, MA, 1991)*. Contemporary Mathematics, vol. 165 (American Mathematical Society, Providence, 1994), pp. 183–211
8. R. Greenberg, Trivial zeros of p-adic L-functions, *p-Adic Monodromy and the Birch and Swinnerton-Dyer Conjecture (Boston, MA, 1991)*. Contemporary Mathematics, vol. 165 (American Mathematical Society, Providence, 1994), pp. 149–174
9. R. Greenberg, V. Vatsal, Iwasawa invariants of elliptic curves. Invent. Math. **142**(1), 17–63 (2000)
10. H. Hida, Galois representations into $\mathrm{GL}_2(Z_p[[X]])$ attached to ordinary cusp forms. Invent. Math. **85**(3), 545–613 (1986)
11. H. Hida, Greenberg's $\mathcal{L}$-invariants of adjoint square Galois representations. Int. Math. Res. Not. **59**, 3177–3189 (2004)
12. H. Hida, *Hilbert Modular Forms and Iwasawa Theory*. Oxford Mathematical Monographs (The Clarendon Press, Oxford University Press, Oxford, 2006)
13. H. Hida, *Elliptic Curves and Arithmetic Invariants*. Springer Monographs in Mathematics (Springer, New York, 2013)
14. K. Kato, M. Kurihara, T. Tsuji, Local Iwasawa theory of Perrin-Riou and syntomic complexes (1996)
15. S. Kobayashi, An elementary proof of the Mazur-Tate-Teitelbaum conjecture for elliptic curves. Doc. Math. 567–575 (2006) (Extra Volume)
16. K. Kitagawa, On standard p-adic L-functions of families of elliptic cusp forms, *p-Adic Monodromy and the Birch and Swinnerton-Dyer Conjecture (Boston, MA, 1991)*. Contemporary Mathematics, vol. 165 (American Mathematical Society, Providence, 1994), pp. 81–110
17. C.P. Mok, $\mathcal{L}$-invariant of the adjoint Galois representation of modular forms of finite slope. J. Lond. Math. Soc. (2) **86**(2), 626–640 (2012)
18. B. Mazur, P. Swinnerton-Dyer, Arithmetic of Weil curves. Invent. Math. **25**, 1–61 (1974)
19. B. Mazur, J. Tate, J. Teitelbaum, On p-adic analogues of the conjectures of Birch and Swinnerton-Dyer. Invent. Math. **84**(1), 1–48 (1986)
20. G. Rosso, A formula for the derivative of the p-adic L-function of the symmetric square of a finite slope modular form. Am. J. Math. **138**(3), 821–878 (2016)
21. A. Wiles, Modular elliptic curves and Fermat's last theorem. Ann. Math. (2) **141**(3), 443–551 (1995)

Quadratic Twists of Elliptic Curves

Yongxiong Li

Abstract In this paper, we give the method of constructing non-torsion points on elliptic curves, which generalizes the classical Birch lemma. As an application, we get more quadratic twist families of the elliptic curve $X_0(49)$, which have rank one. This report is a combination of the two joint works (Coates, Li, Tian, Zhai, Proc Lond Math Soc 110(2), 357–394, 2015, [4]; Cai, Li, Wang, Sci China Math 59(7), 1307–1326, 2016, [2]).

2010 Mathematics Subject Classification 11G05 · 11G40

1 Background

Let $E/\mathbb{Q}$ be an elliptic curve, write in the following form

$$y^2 = x^3 + ax + b, \qquad a, b \in \mathbb{Q}.$$

For any square free rational number $d \in \mathbb{Q}^\times$, define the quadratic twist $E^{(d)}$ of E by the field extension $\mathbb{Q}(\sqrt{d})/\mathbb{Q}$ as follows

$$E^{(d)} : dy^2 = x^3 + ax + b.$$

There is a widely open conjecture for the quadratic twists of elliptic curves defined over $\mathbb{Q}$.

Conjecture 1.1 (D. Goldfeld) *For any elliptic curve E defined over $\mathbb{Q}$, in all its quadratic twist families*

$$\{E^{(d)} | d \in \mathbb{Q}^\times \textit{ square free}\}.$$

Y. Li (✉)
Yau Mathematical Sciences Center, Tsinghua University, Beijing, China
e-mail: liyx_1029@tsinghua.edu.cn

Z. Liang and C. Aribam (eds.), *The Computational and Theoretical Aspects of Elliptic Curves*, Mathematical Lectures from Peking University,
https://doi.org/10.1007/978-981-13-6664-2_4

- *there exists* 50% $d \in \mathbb{Q}^\times/(\mathbb{Q}^\times)^2$, *such that* $\mathrm{rank}_{\mathbb{Z}} E^{(d)}(\mathbb{Q}) = 0$;
- *there exists* 50% $d \in \mathbb{Q}^\times/(\mathbb{Q}^\times)^2$, *such that* $\mathrm{rank}_{\mathbb{Z}} E^{(d)}(\mathbb{Q}) = 1$.

This conjecture is not known even for a special single elliptic curve defined over $\mathbb{Q}$. For the progress on such a conjecture, see the work of Smith [9]. The goal of this short note is given some recent results on the explicit construction of rank one(resp. rank zero) quadratic twist families for elliptic curve defined over $\mathbb{Q}$.

Example 1.2 (*Congruent number problem*) We call a square-free nonzero rational number n is a congruent number if n is equal to the area of a right triangle with its three sides are all rational numbers. A well known fact tell us that n is a congruent number if and only if the rank of the Mordell-Weil group over $\mathbb{Q}$ of the following elliptic curve

$$\mathcal{C}^{(n)} : ny^2 = x^3 - x$$

is positive. Hence, we translate the existence of the congruent number to the study of the quadratic twist of the elliptic curve

$$\mathcal{C} : y^2 = x^3 - x$$

by the field extension $\mathbb{Q}(\sqrt{n})$ over $\mathbb{Q}$.

The results concern the congruent number are given as follows.

Theorem 1.3 (Heegner 1952 [6]) *A square free positive integer n congruent to* $5, 6, 7$ *modulo* 8 *with one odd prime factor is a congruent number.*

Theorem 1.4 (Tian [10]) *For any integer $k \geq 1$, there exists infinitely many square free positive integers congruent to* $5, 6, 7$ *modulo* 8 *with exactly k odd primes factors, which are congruent numbers.*

Now, in the following sections of the paper, we have two goals:

- to study the arithmetic of quadratic twist for general elliptic curves defined over $\mathbb{Q}$(Note that Tian's results only deal with quadratic twist of one elliptic curve defined over $\mathbb{Q}$.);
- generalizing the above method to see some new phenomenon via studying the quadratic twist of the elliptic curve $A := (X_0(49), [\infty])$.

2 Birch Lemma

We first introduce the classical Birch lemma to show the non-torsion of the Heegner points for elliptic curves parametrized by modular curves.

Theorem 2.1 (Birch Lemma) *Let E be an elliptic curve defined over $\mathbb{Q}$ of conductor N, and*

$$f : X_0(N) \to E$$

be the parametrized map sending $[\infty]$ to 0, let $K = \mathbb{Q}(\sqrt{-\ell_0})$ be an imaginary quadratic field.

Assume

(1) Heegner Hypothesis: For every prime factor $\ell \mid N$, ℓ splits in K;
(2) $f([0]) \notin 2E(\mathbb{Q})$;
(3) $\ell_0 \equiv 3 \bmod 4$ is a prime.

Then we have

$$\mathrm{rank}_{\mathbb{Z}}(E^{(-\ell_0)}(\mathbb{Q})) = 1 = \mathrm{ord}_{s=1} L(E^{(-\ell_0)}/\mathbb{Q}, s).$$

Remark 2.2 • Historically, Birch showed the Heegner point is nontrivial, we rewrite his lemma in the above form using the later work of Gross-Zagier and Kolyvagin.

• We explain the conditions which are stated in the Birch lemma, the first is used to construct a Heegner point on the modular curve $X_0(N)$, the second condition comes from the new form theory, which says that for the new form f of conductor N, the Atkin-Lehner operator W_N acts on f via multiplication by $-\epsilon(E/\mathbb{Q})$(Here, we can see all the parametrization maps form a representation space, and our f is a new form in such a representation space, for more precise information on such issue, see for example, [5, 11]), where $\epsilon(E/\mathbb{Q})$ is the root number of $E/\mathbb{Q}$, should be ± 1. By Heegner hypothesis, one can show that $\epsilon(E/\mathbb{Q}) = 1$, since otherwise, one may use the fixed point Q of W_N on upper half plane to show

$$f^{W_N}(Q) - f(Q) := f(W_N.Q) - f(Q) = 0 \in 2E(\mathbb{Q}),$$

which is a contradiction. Therefore, one always has $f(P) + f(P^{W_N}) =$ const for all points P on $X_0(N)$, so, the second condition is to say that this constant point should not in $2E(\mathbb{Q})$. The last condition is just to ensure the class number of K is odd.

For the proof of the Birch lemma. We define the Heegner point,

$$P := (\mathbb{C}/\mathcal{O}_K \to \mathbb{C}/\mathfrak{n}^{-1}) \in X_0(N)(H_K)$$

where H_K is the Hilbert class field of K, and $N\mathcal{O}_K = \mathfrak{n}\bar{\mathfrak{n}}$ is an ideal decomposition of N in K, by condition one, we know that $\mathfrak{n} \neq \bar{\mathfrak{n}}$. Then we will show that the point

$$y_K := \mathrm{Tr}_{H_K/K}(f(P)) \in E(K)$$

belongs to $E(K)^-$ and is non-torsion, then by the following identification

$$E^{(-\ell_0)}(\mathbb{Q}) \simeq E(K)^-,$$

we get the results. Here, we denote by $E(K)^-$ the subgroup of $E(K)$ consists of points which acts by -1 under the non-trivial element in the Galois group $\mathrm{Gal}(K/\mathbb{Q})$.

We remark here a relation of the action of the Atkin-Lehner operator W_N and complex conjugation τ,

$$P^{W_N} = P^{\tau\sigma(\mathfrak{n})}, \tag{2.1}$$

where $\sigma(\mathfrak{n})$ denotes the Artin symbol of $\mathfrak{n}$ in $\mathrm{Gal}(K^{\mathrm{ab}}/K)$.

Proof of Theorem (2.1) By the above remark, we know that

$$f(P) + f(P^{W_N}) = f([0]) \notin 2E(\mathbb{Q}), \tag{2.2}$$

We can see easily that $f([0])$ is of even order, thus via multiplying a suitable odd positive integer, we may assume that $f([0])$ is of order 2^δ with $\delta \geq 1$. We take trace of the equality (2.2) from H_K to K, and notice the relation (2.1), then we have

$$y_K + y_K^\tau = [H_K : K] \cdot f([0]) \in E(\mathbb{Q})[2^\delta]. \tag{2.3}$$

Note that $[H_K : K]$ is an odd integer, so $[H_K : K]f([0]) \notin 2E(\mathbb{Q})$. Moreover, we know that

$$2^\delta y_K \in E(K)^-.$$

Now, we claim that it is non-torsion, we argue by contradiction. From $(\ell_0, 2N) = 1$, thus by the ramification consideration, we know that

$$2^\delta y_K \in E(K)[2^\infty] = E(\mathbb{Q})[2^\infty],$$

also

$$y_K \in E(\mathbb{Q})[2^\infty].$$

Thus

$$y_K + y_K^\tau = 2y_K \in 2E(\mathbb{Q})$$

is a contradiction to (2.3). Therefore, we know that y_K is a non-torsion point. □

3 Generalized Birch Lemma and Rank One Twists

For an ablian group W, denote by $\widehat{W} = W \otimes_{\mathbb{Z}} \widehat{\mathbb{Z}}$.

Let B be an indefinite quaternion algebra over $\mathbb{Q}$ with discriminant d_B and view $B^\times$ as a subgroup of $\mathrm{GL}_2(\mathbb{R})$ via an isomorphism $B \otimes_{\mathbb{Q}} \mathbb{R} \cong M_2(\mathbb{R})$, denote by $B_+^\times$ the set of elements with positive determinant in $B^\times$, for any $\mathbb{Q}$-algebra L, we write

B_L the base change of B to L. There is a projective system of compact Riemann surfaces

$$X_U(\mathbb{C}) \cong B_+^\times \backslash (\mathcal{H} \cup \mathbb{P}^1) \times B_{\mathbb{A}_f}^\times / U,$$

indexed by open compact subgroups U of $B_{\mathbb{A}_f}^\times$ and with connected map

$$\varphi_{UU'} : X_U(\mathbb{C}) \to X_{U'}(\mathbb{C}), \qquad U \subset U'.$$

In the following, for any point $P \in X_U(\mathbb{C})$, we denote by $[h, g]$ with $h \in \mathcal{H} \cup \mathbb{P}^1$, $g \in B_{\mathbb{A}_f}^\times$ for the point P.

This system has a canonical descent to the projective system of algebraic curves over $\mathbb{Q}$, which is projective smooth and irreducible, but not necessarily geometrically irreducible algebraic curve over $\mathbb{Q}$.

We have the Hecke action on X_U, it is defined as follows, for any $t \in B_{\mathbb{A}_f}^\times$, it maps $X_U(\mathbb{C}) \longrightarrow X_{t^{-1}Ut}(\mathbb{C})$, which on the points is given by $[h, g] \mapsto [h, gt]$. It is well known that the Hecke action of the right multiplication by t also descents to $\mathbb{Q}$.

For the Shimura curve X_U, we know that for each element t in the normalizer $\mathrm{N}_{B_{\mathbb{A}_f}^\times}(U)$ of U in $B_{\mathbb{A}_f}^\times$, the Hecke action of right multiplication by t on X_U gives an automorphism on the curve, it is known that the Hecke action is defined over $\mathbb{Q}$. Thus, we have the following map,

$$\mathrm{N}_{B_{\mathbb{A}_f}^\times}(U) \to \mathrm{Aut}_{\mathbb{Q}}(X_U).$$

Let K be an imaginary quadratic field with discriminant $D < 0$, such that there exists an $\mathbb{Q}$-algebra embedding

$$K \to B.$$

Denote by K^- the K-module of elements $j \in B$ such that $jt = \bar{t}j$ for all $t \in K$, where $t \mapsto \bar{t}$ is the non-trivial element in $\mathrm{Gal}(K/\mathbb{Q})$. Then we give the following definition.

Definition 3.1 Let X_U be a Shimura curve, $K \subset B$ be an imaginary quadratic field which is embedded into B. We call an automorphism w on Shimura curve X_U a **special automorphism** for the pair (X_U, K), if w can be written as $w = tj$, where $t \in \widehat{K}^\times$, $j \in K^-$, up to right multiplication by an element in U.

Example 3.2 • For the pair $(X_0(N), K)$, and assume that for any prime divisor $\ell \mid N$, ℓ splits in K, we know that the element

$$W_N = \begin{pmatrix} 0 & 1 \\ -N & 0 \end{pmatrix}$$

is a special automorphism for $(X_0(N), K)$.

• Let $K = \mathbb{Q}(\sqrt{-\ell_0})$, $\ell_0 \equiv 3 \mod 4$, $\ell_0 > 0$ be a prime, $B = \mathrm{GL}_2$ over $\mathbb{Q}$. Let $R \subset \mathcal{O}_B$, which is an order of the maximal order $\mathcal{O}_B$ with conductor $N_-^2 N_+$, where

N_- is the product of inert primes in K and N_+ is the product of split primes in K. Take $U = \widehat{R}^\times$. Embed K into B via the following map,

$$\sqrt{-\ell_0} \mapsto \begin{pmatrix} a & -2 \\ \frac{\ell_0+a^2}{2} & -a \end{pmatrix} \qquad a^2 \equiv -\ell_0 \mod 4N_+.$$

Then the special automorphism exists, for details of the construction of the automorphism, see [1].

Now, we state the generalized Birch lemma in the following

Theorem 3.3 (Generalized Birch Lemma) *Let E be an elliptic curve over $\mathbb{Q}$ of conductor N and let $f : X_U \to E$ be a modular parametrization by a Shimura curve associated to an indefinite quaternion algebra B with level $U \supset \widehat{\mathbb{Z}}^\times$. Let $K \subset B$ be an imaginary quadratic field of discriminant $D \neq -3, -4$.*

Assume that

(1) there is an element $w \in N_{\widehat{B}^\times} U$, which is a special automorphism for (X_U, K), and such that $f + f^w$ is a constant map which values on a torsion point $Q \notin 2E(\mathbb{Q})$.
(2) $(D, 2N) = 1$ and $[H : K]$ is odd, where H is the abelian extension over K such that $\mathrm{Gal}(H/K) \cong \widehat{K}^\times / K^\times (U \cap \widehat{K}^\times)$ under the reciprocity law in class field theory.

Then

$$\mathrm{rank}_{\mathbb{Z}}(E^{(D)}(\mathbb{Q})) = 1 = \mathrm{ord}_{s=1}(L(E^{(D)}/\mathbb{Q}, s))$$

Using the above lemma, combining with the Euler system properties of Heegner points over the towers of ring class field over K, we can state the main theorem of this section in the following.

First,we give some notations.

- Let $E/\mathbb{Q}$ be an elliptic curve of conductor N, which is parametrized by Shimura curve $X_U/\mathbb{Q}$, the Shimura curve associates to an indefinite quaternion algebra B defined $\mathbb{Q}$ of level U with $U \supset \widehat{\mathbb{Z}}^\times$. Denote by $f : X_U \to E$ the defined over $\mathbb{Q}$ morphism,
- Let $\phi = \sum_{n=1}^{\infty} a_n q^n$ be the associate weight 2 newform with respect to E,
- Let K be an imaginary quadratic field with discriminant D, with an embedding $K \hookrightarrow B$.
- Let $S = S_U$ be a set of finite places of $\mathbb{Q}$ containing all places dividing $2ND$ such that $U = U_S U^S$ with $U_S \subset \prod_{v \in S} B(\mathbb{Q}_v)$ and U^S is a maximal open compact subgroup of $B(\mathbb{A}_f^S)$. Here, $\mathbb{A}_f^S$ is denoted by the S-off finite idéle of $\mathbb{Q}$.

We denote by σ_t^K the Artin symbol of $t \in \widehat{K}^\times$ in $\mathrm{Gal}(K^{\mathrm{ab}}/K)$.

Theorem 3.4 *Notations given as above, denote $H = H_U$ the abelian extension over K corresponding to the class group $K^\times(U \cap \widehat{K}^\times)$, assume*

(1) $E(\mathbb{Q})$ has no order 4 torsion points.
(2) $(D, 2N) = 1$ and $[H : K]$ is odd.
(3) there exists $w = t_0 j \in N_{\widehat{B}^\times} U$ which is a special automorphism for (E, K) such that the morphism

$$f + f^w : X_U \to E$$

is constant and valued at a torsion point $Q_0 \notin 2E(\mathbb{Q})$.

For any $r \geq 1$, let Σ_r denote the set of primes $\ell \notin S$ satisfying:

(1) $a_\ell \equiv 0 \mod 2^{r+1}$,
(2) $\ell \equiv 1 \mod 4$,
(3) $\sigma_{t_0}^K(\sqrt{\ell}) = \sqrt{\ell}$

Then for any integer $M = \ell_1 \cdots \ell_r$ with $\ell_1, \cdots, \ell_r \in \Sigma_r$,

$$\mathrm{rank}_{\mathbb{Z}}(E^{(DM)}(\mathbb{Q})) = 1 = \mathrm{ord}_{s=1}(L(E^{(DM)}/\mathbb{Q}, s))$$

and

$$\mathrm{rank}_{\mathbb{Z}}(E^{(M)}(\mathbb{Q})) = 0 = \mathrm{ord}_{s=1}(L(E^{(M)}/\mathbb{Q}, s))$$

Remark 3.5 The cardinality of Σ_r could be finite(even empty!). But the below corollary shows that there are many cases that this cardinality could be infinity.

By the Chebotarev theorem, we get the following corollary.

Corollary 3.6 *Let $f : X_U \to E$ a modular parametrization of an elliptic curve over $\mathbb{Q}$ by a Shimura curve associated to a quaternion algebra B. Let $K \subset B$ be an imaginary quadratic field of discriminant D. Assume the conditions* (1), (2), *and* (3) *in Theorem 3.4 and the following (4) there is a supersingular good prime q for E with $q \equiv 1 \mod 4$ and $\sigma_{t_0}^K(\sqrt{q}) = \sqrt{q}$.*

Then for any integer $k \geq 1$, there are infinitely many square-free M with exactly k odd prime factors such that

$$\mathrm{ord}_{s=1} L(E^{(M)}, s) = 0 \quad \text{and} \quad \mathrm{ord}_{s=1} L(E^{(DM)}, s) = 1.$$

Example 3.7 Let $A = X_0(49)$, $K = \mathbb{Q}(\sqrt{-\ell_0})$, $\ell_0 \equiv 3 \mod 4$, $\ell_0 > 3$ a prime number, and let

$$R = q_1 \cdots q_r,$$

where $q_i \equiv 1 \mod 4$ inert in $\mathbb{Q}(\sqrt{-7})$, then we have

$$\mathrm{rank}_{\mathbb{Z}}(A^{(-\ell_0 R)}(\mathbb{Q})) = 1.$$

4 Some New Phenomenon for Quadratic Twists of $A = X_0(49)$

For the curve $A = X_0(49)$, by using the theorem which is obtained in the last section, and by combining the follow three ingredients.

- The Gross-Zagier formula in the explicit form of Cai-Shu-Tian [3].
- The induction argument of Heegner points of Tian [10].
- The work of Perrin-Riou [8] and Kobayashi [7].

We state the final theorem of this note in the following.

Theorem 4.1 ([2, 4]) *Let* $A = (X_0(49), [\infty])$, $M = -\ell_0 RN$, *satisfying*

- *the positive prime* ℓ_0 *is greater than* 3, *which is a non-quadratic modulo* 7.
- *the product* $R = q_1 \cdots q_r$, *with* $q_i (1 \leq i \leq r)$ *is congruent to* 1 *modulo* 4, *which is inert in* $\mathbb{Q}(\sqrt{-7})$.
- *the product* $N = p_1 \cdots p_k$, *with* $p_j (1 \leq j \leq k)$ *is congruent to* 1 *modulo* 4, *which is split completely in* $\mathbb{Q}(A[4], \sqrt{R})$.

Assume moreover that

$$\text{the class group of } \mathbb{Q}(\sqrt{-\ell_0 N}) \text{ has no element of order } 4.$$

Then

$$\operatorname{rank}_{\mathbb{Z}}(A^{(M)}(\mathbb{Q})) = 1 = \operatorname{ord}_{s=1} L(A^{(M)}, s),$$

and the ℓ*-part B-SD conjecture holds for* $A^{(M)}$ *for all primes* $\ell \nmid (7M)$.

Acknowledgements The author would like to thank Ye Tian to give him a chance to present the joint work in the conference which was hold at Beijing university in the year 2014. The author would also like to thank the organizers of that conference for their hospitality.

References

1. L. Cai, Y. Chen, Y. Liu, Heegner points on modular curves. Trans. Am. Math. Soc. **370**(5), 3721–3743 (2018)
2. L. Cai, Y. Li, Z. Wang, Special automorphism on Shimura curves and non triviality of Heegner points. Sci. China Math. **59**(7), 1307–1326 (2016)
3. L. Cai, J. Shu, Y. Tian, Explicit Gross-Zagier and Waldspurger formulae. Algebr. Number Theory **8**(10), 2523–2572 (2014)
4. J. Coates, Y. Li, Y. Tian, S. Zhai, Quadratic twists of elliptic curves. Proc. Lond. Math. Soc. (3) **110**(2), 357–394 (2015)
5. B. Gross, *Heegner Points and Representation Theory*. Heegner Points and Rankin L-Series, vol. 49 (Mathematical Sciences Research Institute Publications, Cambridge University Press, Cambridge, 2004), pp. 37–65
6. K. Heegner, Diophantische analysis und modulfunktionen. Math. Z. **56**, 227–253 (1952)

7. S. Kobayashi, The p-adic Gross-Zagier formula for elliptic curves at supersingular primes. Invent. Math. **191**(3), 527–629 (2013)
8. B. Perrin-Riou, Points de Heegner et derivees de fonctions L p-adiques. Invent. Math. **89**(3), 455–510 (1987)
9. A. Smith, 2^{∞}-Selmer groups, 2^{∞}-class groups, and Goldfelds conjecture, arXiv:1702.02325v1
10. Y. Tian, Congruent numbers and Heegner points. Camb. J. Math. **2**(1), 117–161 (2014)
11. X. Yuan, S. Zhang, W. Zhang, *The Gross-Zagier Formula on Shimura Curves*. Annals of Mathematics Studies, vol. 184 (Princeton University Press, Princeton, 2013), x+256 pp. ISBN: 978-0-691-15592-0

Computing Fourier Coefficients of Level One Modular Forms

Peng Tian

Abstract In this paper we describe an algorithm to compute the Fourier coefficients of level one modular forms f. We then give an example to show how to do the explicit computations. We also show how to use the results to find a bound B of primes p such that the Fourier coefficients $a_p(f) \neq 0$ for all $p < B$. As examples, we compute the explicit bounds B_k for the unique cusp forms Δ_k of level one and weight k with $k = 16, 18, 20, 22, 26$.

2010 Mathematics Subject Classification 11-04 · 11Fxx · 11G30 · 11Rxx · 11Y40 · 11Y55

1 Introduction

For large prime p, it is an interesting problem to compute Ramanujan's tau function $\tau(p)$ defined by

$$\Delta(z) = q \prod_{n \geq 1} (1 - q^n)^{24} = \sum_n \tau(n) q^n.$$

In the book [6], S.J. Edixhoven, J.-M. Couveignes, R.S. de Jong and F. Merkl generalize Schoof's algorithm [11] and show that

There exists a deterministic algorithm that on input a prime number p *computes* $\tau(p)$ *in time polynomial in* $\log p$.

Ramanujan observed the property of $\tau(p)$:

$$|\tau(p)| \leq 2p^{11/2} \text{ for prime } p,$$

P. Tian (✉)
Department of Mathematics, East China University of Science and Technology, Shanghai 200237, People's Republic of China
e-mail: tianpeng135@163.com

Z. Liang and C. Aribam (eds.), *The Computational and Theoretical Aspects of Elliptic Curves*, Mathematical Lectures from Peking University,
https://doi.org/10.1007/978-981-13-6664-2_5

which was proved by P. Deligne. In fact, he [3] shows that there exists a continuous semi-simple representation

$$\rho_{\Delta,\ell} : Gal(\overline{\mathbb{Q}}|\mathbb{Q}) \to GL_2(\overline{\mathbb{F}}_\ell).$$

Moreover, this representation has the property that for primes p not dividing $N\ell$ one has

$$\tau(p) \equiv \mathrm{tr}(\rho_{\Delta,\ell}(\mathrm{Frob}_p)) \mod \ell.$$

In [6], Edixhoven and Couveignes give a polynomial time algorithm to compute the modular Galois representation and thus the value modulo ℓ of Ramanujan's tau function at p. Then combining with the property $|\tau(p)| \leq 2p^{11/2}$ for primes p and the Chinese remainder theorem one can compute $\tau(p)$.

Unfortunately the algorithm described in [6] is difficult to implement. Bosman [1] used this algorithm to approximately evaluate $\tilde{P}_{f,\ell}$ of mod ℓ Galois representations associated to modular forms f of level 1 and of weight $k \leq 22$, with $\ell \leq 23$. In [17] P. Tian presented an improvement in case $\gcd(k-2, l+1) > 2$. He worked with the Jacobian of X_Γ rather than the Jacobian of $X_1(\ell)$ that Bosman used. Since the genus of X_Γ is smaller than that of $X_1(\ell)$, the required precision in the computations is smaller and the computation is more efficient. This allows him to deal with cases that were inaccessible by Bosman's original algorithm. He in fact explicitly computed the case with $\ell = 29$ and f of weight $k = 16$, and the cases with $\ell = 31$ and f of weight $k = 12, 20, 22$.

In this paper we describe the improved algorithm and give an example to show how to do the explicit computations.

We also show in this paper that the modular polynomials $\tilde{P}_{f,\ell}$, together with the congruent formulations of $a_p(f)$ modulo exceptional primes, can be used to calculate a bound of primes p for which $a_p(f)$ is non-vanishing. In fact for a prime number $p \neq \ell$, we can verify $a_p(f) \equiv 0 \bmod \ell$ by checking whether the projective modular polynomial $\tilde{P}_{f,\ell}$ has an irreducible factor of degree 2 over $\mathbb{F}_p$.

As examples, we achieve the explicit bounds B_k for the unique cusp form Δ_k of level one and weight k with $k = 16, 18, 20, 22, 26$ such that $a_n(\Delta_k) \neq 0$ for all $n < B_k$.

2 An Algorithm in Polynomial Time

Let N be a positive integer. The congruence subgroup $\Gamma_1(N)$ of level N is

$$\Gamma_1(N) = \left\{ \begin{pmatrix} a & b \\ c & d \end{pmatrix} \in SL(2,\mathbb{Z}) \mid c \equiv 0 \; mod \; N, \quad a \equiv b \equiv 1 \; mod \; N \right\}.$$

Let $k \geq 2$ be an integer. Let $f = \sum_{n>0} a_n(f)q^n \in S_k(\Gamma_1(N))$ be a newform of weight k and level N. Let ε be its nebentypus character. Let K_f be the number field which is obtained by adjoining all coefficients a_n of the q-expansion f to $\mathbb{Q}$. Let ℓ be a prime number. Let λ be a prime of K_f lying over ℓ. Then we have the following well known theorem:

Theorem 2.1 (Deligne) *$f \in S_k(N, \varepsilon)$ be a newform. Let λ be as above and let $\mathbb{F}_\lambda$ denote the residue field of λ. Then there exists a continuous semi-simple representation*

$$\rho_{f,\lambda} : Gal(\overline{\mathbb{Q}}|\mathbb{Q}) \to GL_2(\mathbb{F}_\lambda).$$

that is unramified outside $N\ell$, and for all primes $p \nmid N\ell$ the characteristic polynomial of $\rho_{f,\lambda}(Frob_p)$ satisfies

$$charpol(\rho_{f,\lambda}(Frob_p)) \equiv x^2 - a_p(f)x + \varepsilon(p)p^{k-1} \mod \lambda. \tag{2.1}$$

Moreover, $\rho_{f,\lambda}$ is unique up to isomorphism.

Since $\mathbb{F}_\lambda \subset \overline{\mathbb{F}}_\ell$, we can view our representation $\rho_{f,\lambda}$ as taking values in $GL_2(\overline{\mathbb{F}}_\ell)$. The discriminant modular form is given by

$$\Delta(z) = q \prod_{n \geq 1} (1 - q^n)^{24} = \sum_n \tau(n) q^n,$$

and its Fourier coefficients define the Ramanujan's tau function $\tau(n)$.

In the book [6], S. Edixhoven and J.-M. Couveignes show that

There exists a deterministic algorithm that computes the mod ℓ *Galois representation associated to level one modular forms in time polynomial in ℓ.*

By computing the mod ℓ Galois representation, we mean to give a basis of the field $K_\ell = \overline{\mathbb{Q}}^{\mathrm{Ker}\rho_{f,\lambda}}$ over $\mathbb{Q}$ and an injective morphism form $\mathrm{Gal}(K_\ell|\mathbb{Q})$ to $GL_2(\mathbb{F}_\lambda)$.

Since we have the congruence relation

$$\tau(p) \equiv \mathrm{tr}(\rho_{\Delta,\ell}(\mathrm{Frob}_p)) \mod \ell,$$

their algorithm can be used to compute $\tau(p)$ mod ℓ in time polynomial in $\log p$ and ℓ.

Fix a prime number ℓ and let λ be a prime lying over ℓ. The following theorem allows Edixhoven and Couveignes to reduce the questions to the weight 2 cases.

Theorem 2.2 (Serre, Gross) *Let $f \in S_k(\Gamma_1(N))$ be a newform. If $2 < k \leq \ell + 1$ and $\rho_{f,\lambda}$ is irreducible, then there is a newform $f_2 \in S_2(\Gamma_1(N\ell))$, together with a prime λ_2 lying over ℓ of the coefficient field K_{f_2}, such that ρ_{f_2,λ_2} is isomorphic to $\rho_{f,\lambda}$.*

Now we suppose that $\rho_{f,\lambda}$ is a mod ℓ Galois representation associated to a newform $f \in S_2(\Gamma_1(\ell))$ with character ε. Denoting $\mathbb{T}$ the Hecke algebra generated by the diamond and Hecke operators over $\mathbb{Z}$, i.e.

$$\mathbb{T} = \mathbb{Z}[T_n, \langle n \rangle : n \in \mathbb{Z}_+ \text{ and } (n, \ell) = 1].$$

Define a ring homomorphism as follows

$$\theta : \mathbb{T} \longrightarrow \mathbb{F}_\ell, \quad \langle d \rangle \mapsto \varepsilon(d), \quad T_n \mapsto a_n(f).$$

Let $\mathfrak{m}$ denote the kernel of θ and put

$$V_\lambda = J_1(\ell)(\overline{\mathbb{Q}})[\mathfrak{m}] = \{x \in J_1(\ell)(\overline{\mathbb{Q}}) \mid tx = 0 \text{ for all } t \text{ in } \mathfrak{m}\},$$

where $J_1(\ell)$ is the Jacobian variety of the modular curve $X_1(\ell)$ associated to $\Gamma_1(\ell)$. This is a 2-dimensional $\mathbb{T}/\mathfrak{m}$-linear subspace of $J_1(\ell)(\overline{\mathbb{Q}})[\ell]$ and the semisimplification of the representation

$$\rho : Gal(\overline{\mathbb{Q}}/\mathbb{Q}) \to \mathrm{Aut}(V_\lambda)$$

is isomorphic to $\rho_{f,\lambda}$ (see [10, Sects. 3.2 and 3.3]).

If $\# \ \mathbb{T}/\mathfrak{m} = \ell$, then the fixed field of $\rho_{f,\lambda}$ is naturally the splitting field of a suitable polynomial $P_{f,\lambda} \in \mathbb{Q}[X]$ of degree $\ell^2 - 1$. In fact, we can write each divisor $P \in V_\lambda - \{0\}$ as $\sum_{i=1}^{g}(P_i) - gO)$ for certain points P_i on $X_1(\ell)$, where g is the genus of $X_1(\ell)$. Then for some suitable function h in the function field of $X_1(\ell)$, we define $h(P)$ as

$$h(P) = \sum_{i=1}^{g} h(P_i).$$

Then we can take

$$P_{f,\lambda}(x) = \prod_{P \in V_\lambda - \{0\}} (x - h(P)). \tag{2.2}$$

3 An Improved Algorithm

As explained in Sect. 2, our main task is then to compute the 2-dimensional $\mathbb{F}_\lambda$-linear space V_λ. We can always do the computations with the Jacobian variety of modular curve $X_1(\ell)$ which has genus $(\ell - 5)(\ell - 7)/24$. However, in this section we will show a method which allows us to work with a modular curve that sometimes has smaller genus than $X_1(\ell)$.

3.1 Finding Modular Curves

Let $k > 0$ be an even integer and let ℓ be a prime number with $k \leq \ell + 1$. Let $f \in S_k(SL(2, \mathbb{Z}))$ be a newform of level 1. Then we have

Theorem 3.1 ([17, Prop 4.1]) *Let $k > 0$ be an even integer and $f \in S_k(SL(2,\mathbb{Z}))$ be a newform of level 1 and weight k. Let $\ell \geq k-1$ be a prime number and λ be a prime lying over ℓ. Let Γ be the unique group*

$$\Gamma_1(\ell) \subset \Gamma \subset \Gamma_0(\ell)$$

with $[\Gamma : \Gamma_1(\ell)] = \frac{1}{2}gcd(k-2, \ell-1)$. Then there exists a newform $f_2 \in S_2(\Gamma)$ and a prime λ_2 lying over ℓ in the field K_{f_2} such that $\rho_{f,\lambda}$ is isomorphic to ρ_{f_2,λ_2}.

Proof It follows from [9, Theorem 2.2] that there exists $f_2 \in S_2(\Gamma_1(\ell))$ and a prime $\lambda_2|\ell$ such that $\rho_{f,\lambda}$ is isomorphic to ρ_{f_2,λ_2}. Since the character of f is trivial in our case, for any $p \nmid \ell$, by (2.1) we have the equalities in $\overline{\mathbb{F}}$:

$$a_p(f_2) = a_p(f) \quad and \quad \varepsilon_2(p) = p^{k-2} \tag{3.1}$$

Here ε_2 is the nebentypus character of f_2, which is a Dirichlet character of the cyclic group $(\mathbb{Z}/\ell\mathbb{Z})^*$. Let ω be the cyclotomic character and then it follows from the second equation in (3.1) that $\varepsilon_2 = \omega^{k-2}$.

By the Galois theory of function fields of modular curves, we know the extension $\mathbb{C}(X(\ell))|\mathbb{C}(X(1))$ is Galois with Galois group

$$\mathrm{Gal}(\mathbb{C}(X(\ell))|\mathbb{C}(X(1))) \cong SL_2(\mathbb{Z}/\ell\mathbb{Z})/\{\pm I\}.$$

Moreover, we have the following 1-1 correspondence for a congruence subgroup Γ of level ℓ

$$\begin{array}{ccc} \textit{Galois extension fields} & \leftrightarrow & \textit{subgroups } \{\pm I\}\Gamma/\{\pm I\}\Gamma(\ell) \\ \mathbb{C}(X_\Gamma) \textit{ of } \mathbb{C}(X(1)) & & \textit{of } SL_2(\mathbb{Z})/\{\pm I\}\Gamma(\ell) \end{array}$$

Now let H denote the normal subgroup $\ker(\omega^{k-2})/\{\pm 1\}$ of $(\mathbb{Z}/\ell\mathbb{Z})^*/\{\pm 1\}$. Then there exists an intermediate curve X of $X_1(\ell) \to X_0(\ell)$ such that the Galois group of $X_1(\ell) \to X$ is H, ie.

$$\begin{array}{cc} X_1(\ell) & \{1\} \\ \downarrow & \\ X & H \\ \downarrow & \\ X_0(\ell) & (\mathbb{Z}/\ell\mathbb{Z})^*/\{\pm 1\} \end{array}$$

Let φ denote the surjection:

$$\varphi: \Gamma_0(\ell) \twoheadrightarrow (\mathbb{Z}/\ell\mathbb{Z})^*, \qquad \begin{pmatrix} a & b \\ c & d \end{pmatrix} \to \overline{d} \tag{3.2}$$

whose kernel is $\Gamma_1(\ell)$. Let Γ_H be the preimage of $\{\pm 1\}H$ under φ. Then we have $X = X_{\Gamma_H}$ and $\ker(\varphi) \subseteq \Gamma_H$, since $\#H = \frac{1}{2}\gcd(k-2, \ell-1)$.

To complete the proof we only need to check that $f_2 \in S_2(\Gamma_1(\ell))$ also lies in $S_2(\Gamma_H)$. In fact, for all $\gamma = \begin{pmatrix} a & b \\ c & d \end{pmatrix} \in \Gamma_H$, it follows from the definition of Γ_H that $\varphi(\gamma)$ is in $\ker(\omega^{k-2})$ and thus $f_2|_2\gamma = \omega^{k-2}(\varphi(\gamma))f_2 = f_2$, which implies $f_2 \in S_2(\Gamma_H)$. □

3.2 Realization of ρ in the Jacobian of a Modular Curve

Let $k > 0$ be an even integer. Suppose $H = \ker(\omega^{k-2})/\{\pm 1\}$. In this subsection, we assume that $\ell \geq k - 1$ is a prime number and $f \in S_2(\Gamma_1(\ell), \omega^{k-2})$. Then we have

$$f \in S_2(\Gamma_H),$$

where Γ_H corresponds to H via φ in (3.2). Let X_{Γ_H} be the modular curve of the subgroup Γ_H and denote J_{Γ_H} its Jacobian, ie.

$$J_{\Gamma_H} = \Omega^1_{\text{hol}}(X)^\wedge / H_1(X, \mathbb{Z}).$$

Then we have $H = \mathrm{Gal}(\mathbb{C}(X_1(\ell))|\mathbb{C}(X_{\Gamma_H})$.

Since the meromorphic differentials of the modular curve X_{Γ_H} form a 1-dimensional vector space over $\mathbb{C}(X_{\Gamma_H})$ generated by $\mathrm{d}f$ for a non-constant function $f \in \mathbb{C}(X_{\Gamma_H})$ that is Γ_H-invariant, the differentials space is isomorphic to $\mathbb{C}(X_{\Gamma_H})$ as $\mathrm{Gal}(\mathbb{C}(X_{\Gamma_H})|\mathbb{C}(X_0(\ell)))$-module. It follows that the holomorphic differential space $\Omega^1_{hol}(X_{\Gamma_H})$ is the H-invariant part of $\Omega^1_{hol}(X_1(\ell))$. By taking duals of these spaces, we get

$$J_{\Gamma_H}(\overline{\mathbb{Q}})[\ell] = J_1(\ell)(\overline{\mathbb{Q}})[\ell]^H := \{x \in J_1(\ell)(\overline{\mathbb{Q}})[\ell] \mid \sigma(x) = x, \text{ for all } \sigma \in H\}.$$

As discussed in Sect. 2, the representation associated to f is a subrepresentation of the ℓ-torsion points of $J_1(\ell)$. However, one can work with J_{Γ_H} instead of $J_1(\ell)$:

Theorem 3.2 ([17, Prop 4.2]) *The torsion space V_λ is a 2-dimensional subspace of $J_{\Gamma_H}(\overline{\mathbb{Q}})[\ell]$.*

Proof It follows from the definition of H that each $\sigma \in H$ acts on $J_1(\ell)(\overline{\mathbb{Q}})[\ell]$ the same as a diamond operator $\langle d \rangle$ for some $d \in (\mathbb{Z}/\ell\mathbb{Z})^*$ with $d^{k-2} = 1$. This implies that $\sigma - id$ is an element of $\mathfrak{m} = \ker(\theta)$ and thus $V_\lambda \subset J_1(\ell)(\overline{\mathbb{Q}})[\ell]^H = J_{\Gamma_H}(\overline{\mathbb{Q}})[\ell]$. □

3.3 Description of the Computations

First of all, by the isomorphisms

$$J_\Gamma(\mathbb{C})[\ell] \cong H_1(X_\Gamma, \mathbb{F}_\ell) \cong \mathbb{S}_2(\Gamma) \otimes \mathbb{F}_\ell,$$

we can describe $J_\Gamma(\mathbb{C})[\ell]$ in terms of modular symbols. Since the period lattice $\Lambda \subset \mathbb{C}^g$ and the action of $\mathbb{T}' \subset \mathrm{End}(J_\Gamma)$ on $\mathbb{S}_2(\Gamma)$ can be numerically computed [14], we thus can approximately compute the torsion points of $V_\lambda \subset \frac{1}{\ell}\Lambda/\Lambda$. Then using the Newton iteration approximation method, we can find torsion divisors via the Abel–Jacobi map and finally obtain the polynomial

$$P_{f,\lambda}(x) = \prod_{P \in V_\lambda - \{0\}} (x - h(P)).$$

Now we can compute the Galois representation $\rho_{f,\lambda}$. In fact, we first can obtain the splitting field K_ℓ of the computed modular polynomial $P_{f,\lambda}$, which is also the fix field of $\mathrm{Ker}\rho_{f,\lambda}$, ie. $K_\ell = \overline{\mathbb{Q}}^{\mathrm{Ker}\rho_{f,\lambda}}$. The bijection between the roots of $V_\lambda - \{0\}$ and $P_{f,\lambda}$ induces an isomorphism $\mathrm{Gal}(K_\ell|\mathbb{Q}) \cong \mathrm{Im}\rho_{f,\lambda}$ which defines $\rho_{f,\lambda}$.

In order to compute $a_p(f) \mod \ell$, by (2.1) it suffices to compute the conjugacy class of the Frobenius element $(\frac{K_\ell/\mathbb{Q}}{p})$. Take $g(x) \in \mathbb{Z}_\ell[x]$. We first define

$$\mathbb{F}_p[a] = \mathbb{F}_p[x]/(P_{f,\lambda}), \text{ and } \mathrm{Tr}_{\mathbb{F}_p[a]/\mathbb{F}_p} g(a)a^p.$$

For a conjugate class C, the resolvent of $g(x)$ is

$$\Gamma_C(x) = \prod_{\sigma \in C} \left(x - \sum_{i=1}^{\ell+1} g(a_i)\sigma(a_i) \right),$$

where a_i are the roots of $P_{f,\lambda}(x)$. Then we have

$$\left(\frac{K_\ell/\mathbb{Q}}{p}\right) \in C \Longleftrightarrow \Gamma_C(u) \equiv 0 \mod p.$$

See [4] for details.

4 One Example

The projective representation $\tilde{\rho}_{f,\lambda} : Gal(\overline{\mathbb{Q}}|\mathbb{Q}) \to PGL_2(\mathbb{F}_\lambda)$ is defined by $\rho_{f,\lambda}$ composed with the canonical projection map $GL_2(\mathbb{F}_\lambda) \to PGL_2(\mathbb{F}_\lambda)$. Since the projective line $\mathbb{P}(V_\lambda)$ has $\ell + 1$ points, the fixed field of $\tilde{\rho}_{f,\lambda}$ is naturally the splitting field of the polynomial

$$\tilde{P}_{f,\lambda}(x) = \prod_{L \subset \mathbb{P}(V_\lambda)} \left(x - \sum_{P \in L - \{0\}} h(P) \right). \tag{4.1}$$

For $k = 12, 14, 16, 18, 20, 22$ and 26, let Δ_k denote the unique cusp form of level 1 and weight k. In [1], Bosman computed the modular projective polynomials $\tilde{P}_{\Delta_k,\ell}$ for several values of ℓ and k. P. Tian [17] applied the improved algorithm described in Sect. 3 to compute a few more polynomials for the cases of $(k, \ell) = (12, 31)$, $(16, 29)$, $(20, 31)$ and $(22, 31)$. Now we shall take one of the cases as an example, ie. $(k, \ell) = (12, 31)$. Since for this case, $\gcd(k - 2, \ell - 1) > 2$, we can work with J_{Γ_H} which has dimension of 6 rather than $J_1(l)$ of dimension 26.

The modular polynomial is

$$\begin{aligned}\tilde{P}_{\Delta_{12},31} = {} & x^{32} - 4x^{31} - 155x^{28} + 713x^{27} - 2480x^{26} + 9300x^{25} - 5921x^{24} + \\ & 24707x^{23} + 127410x^{22} - 646195x^{21} + 747906x^{20} - 7527575x^{19} \\ & + 4369791x^{18} - 28954961x^{17} - 40645681x^{16} + 66421685x^{15} - \\ & 448568729x^{14} + 751001257x^{13} - 1820871490x^{12} + \\ & 2531110165x^{11} - 4120267319x^{10} + 4554764528x^{9} - \\ & 5462615927x^{8} + 4607500922x^{7} - 4062352344x^{6} + \\ & 2380573824x^{5} - 1492309000x^{4} + 521018178x^{3} - 201167463x^{2} \\ & + 20505628x - 1261963\end{aligned}$$

The polynomial $\tilde{P}_{\Delta_{12},31}$ has also been obtained by Zeng [19]. His method avoids the high precision computations and is based on p-adic computations.

It is difficult to rigorously prove that the computations have been done with sufficient accuracy and that therefore the results are correct. However, one can verify that the computed polynomial is correct using Serre's conjecture which has been fully proved by C. Khare and J.P. Wintenberger in 2008.

Let ℓ be a prime. In [13] Serre defined the a Serre level $N(\rho)$ and a Serre weight $k(\rho)$ for a Galois representation $\rho : Gal(\overline{\mathbb{Q}}|\mathbb{Q}) \to GL_2(\overline{\mathbb{F}}_\ell)$. Edixhoven [5] reformulate the definitions. Then we have

Theorem 4.1 (Serre's Conjecture) *Let ℓ be a prime and let ρ: $Gal(\overline{\mathbb{Q}}|\mathbb{Q}) \to GL_2(\overline{\mathbb{F}}_\ell)$ be a representation that is irreducible and odd. Then there exists a newform f of level $N(\rho)$ and weight $k(\rho)$ and a prime λ of K_f above ℓ such that ρ is isomorphic to $\overline{\rho}_{f,\lambda}$.*

Proof See [7]. □

Now we have

Proposition 4.2 *The polynomial $\tilde{P}_{\Delta_{12},31}$ above is irreducible. The Galois group of its splitting field is isomorphic to $PGL_2(\mathbb{F}_{31})$. Moreover, a subgroup of $Gal(\overline{\mathbb{Q}}|\mathbb{Q})$ fixing a root of $\tilde{P}_{\Delta_{12},31}$ corresponds via $\tilde{\rho}_{\Delta,31}$ to a subgroup of $PGL_2(\mathbb{F}_{31})$ fixing a point of $\mathbb{P}^1(\mathbb{F}_{31})$.*

Sketch of the proof. We denote $K_{12,31} := \mathbb{Q}[x]/(\tilde{P}_{\Delta_{12},31})$ the number field defined by $\tilde{P}_{\Delta_{12},31}$ and the integer ring of $K_{12,31}$ is denoted by $\mathcal{O}_{12,31}$.

Let G be a subgroup of $Gal(\overline{\mathbb{Q}}|\mathbb{Q})$ fixing a root of $\tilde{P}_{\Delta_{12},31}$. Then the group G corresponds to a subgroup of $Gal(\tilde{P}_{\Delta_{12},31})$ of index $\deg(\tilde{P}_{\Delta_{12},31}) = 32$, and thus the image of G via $\tilde{\rho}_{12,31}$ is a subgroup of $PGL_2(\mathbb{F}_{31})$ of index 32.

By the algorithms in [2, Sect. 6], we can show that the discriminant $\mathcal{D}_{12,31}$ of $K_{12,31}$ over $\mathbb{Q}$ is $(-1)^{(\ell-1)/2}\ell^{k+\ell-2} = -31^{41}$. Thus $\tilde{\rho}_{12,31}$ is unramified at all $p \neq 31$. A lemma by Tate implies that $\tilde{\rho}_{12,31}$ has a lifting which is unramified outside 31. Furthermore, the minimal weight of a lifting of $\tilde{\rho}_{k,\ell}$ equals $v_\ell(\mathcal{D}_{k,\ell}) - \ell + 2 = k$. Therefore, $\tilde{\rho}_{12,31}$ has a lifting $\rho_{12,31}$ with weight 12 and level 1.

It is easy to show that the representation $\rho_{12,31}$ is odd and irreducible.

The cuspidal space $S_k(SL_2(\mathbb{Z}))$ has dimension 1. Then Serre's conjecture ensures that $\rho_{12,31} \cong \rho_{\Delta_{12},31}$, and hence $\tilde{\rho}_{12,31} \cong \tilde{\rho}_{\Delta_{12},31}$. □

Then for big primes p, we can compute the following congruences of $\tau(p)$ in $\mathbb{Z}/31\mathbb{Z}$:

$$\begin{aligned}\tau(10^{1000} + 4351) &= \pm 8,\\ \tau(10^{1000} + 10401) &= 0,\\ \tau(10^{1000} + 11979) &= \pm 11,\\ \tau(10^{1000} + 17557) &= \pm 8.\end{aligned}$$

5 Non-vanishing Fourier Coefficients of Level One Modular Forms

In 1947, D.H. Lehmer conjectured that Ramanujan's tau function $\tau(n)$ is non-vanishing for all n. In [8, Theorem 2] he proved that the smallest n for which $\tau(n) = 0$ must be a prime and showed that $\tau(n) \neq 0$ for all $n < 3316799$. In [12], J.-P. Serre showed that if $\tau(p) = 0$ for a prime p, then

$$\begin{aligned}&p \equiv -1 \mod 2^{11}3^7 5^3 691,\\ &p \equiv -1, 19, 31 \mod 7^2 \quad \text{and}\\ &p \equiv \textit{a non-square} \mod 23.\end{aligned}$$

from which he obtained a bound of 15 digits for Lehmer's conjecture with respect to $\tau(n)$.

The modular polynomial can be used to evaluate the Fourier coefficients of modular forms. For a prime number $p \neq \ell$, it can be shown that $a_p(f) \equiv 0 \bmod \ell$ if and only if there exists a prime $\mathfrak{p}|p$ of degree 2 in the number field $K = \mathbb{Q}[x]/(\tilde{P}_{f,\ell})$. Hence, for $p \nmid Disc(\tilde{P}_{f,\ell})$, we can verify $a_p(f) \equiv 0 \bmod \ell$ by checking whether the projective modular polynomial $\tilde{P}_{f,\ell}$ has an irreducible factor of degree 2 over $\mathbb{F}_p$.

This method has been first applied by Bosman [6, Corollary 7.14] to search a bound for Lehmer's conjecture with respect to $\tau(n)$. More precisely, he systematically searched for the smallest prime p in the congruence classes for which in addition $\tau(p) \equiv 0 \mod 11 \cdot 13 \cdot 17 \cdot 19$ and showed that $\tau(n) \neq 0$ for all n with $n < 22798241520242687999$. Zeng [19] and Tian [17] obtained a large bound 982149821766199295999 by adding 31 into the list of primes.

Let Δ_k denote the unique cusp form of level 1 and weight k with $k = 12, 16, 18, 20, 22, 26$. In [15, 16], H.P. F. Swinnerton-Dyer shows a method to determine the exceptional primes ℓ for f and the congruences of $a_p(f)$ modulo the powers of

ℓ. In particular, he explicitly determined almost all the exceptional primes ℓ for Δ_k and for primes $p \neq \ell$ achieved the congruences of $a_p(\Delta_k)$ modulo the powers of ℓ. In fact he showed that there are three types of exceptional primes and for each exceptional prime ℓ of type (i), together with the values of $a_p(f)$ for certain p, one can compute m and $N \geq 1$ in

$$a_p(f) \equiv p^m + p^{k-1-m} \mod \ell^N. \tag{5.1}$$

To summarize Swinnerton-Dyer's results we can obtain the following tables which show the values of m and N in (5.1) for the exceptional primes ℓ type (i) (Tables 1, 2, 3, 4, 5 and 6).

For the case of type (ii), it can be shown that there are two exceptional primes for Δ_k, i.e., $\ell = 23$ when $k = 12$ and $\ell = 31$ when $k = 16$. In fact we also have the explicit congruences in these cases. It is well-known that $a_p(\Delta_{12})$ satisfies the following congruences (see [18]):

$$\begin{aligned} a_p(\Delta_{12}) &\equiv 0 && \mod 23 \; if \; \left(\tfrac{p}{23}\right) = -1, \\ a_p(\Delta_{12}) &\equiv 2 && \mod 23 \; if \; p = u^2 + 23v^2 \; for \; integers \; u \neq 0, v, \\ a_p(\Delta_{12}) &\equiv -1 && \mod 23 \; \; for \; other \; p \neq 23. \end{aligned}$$

Here $(\frac{\cdot}{\cdot})$ is the Legendre symbol.

The results of $\ell = 31$ for Δ_{16} is quite similar and we have

$$\begin{aligned} a_p(\Delta_{16}) &\equiv 0 && \mod 31 \; if \; \left(\tfrac{p}{31}\right) = -1, \\ a_p(\Delta_{16}) &\equiv 2 && \mod 31 \; if \; p = u^2 + 31v^2 \; for \; integers \; u \neq 0, v, \\ a_p(\Delta_{16}) &\equiv -1 && \mod 31 \; \; for \; other \; p \neq 31. \end{aligned}$$

For a prime p, the congruences (5.1) with N and m in Tables 1, 2, 3, 4, 5 and 6 imply: if $a_p(\Delta_{16}) = 0$, then p satisfies

Table 1 k = 16

ℓ	3	5	7	11	3617
N	5 if $(\frac{p}{3}) = 1$ 6 if $(\frac{p}{3}) = -1$	2	3	1	1
m	174	17	85	1	0

Table 2 k = 18

ℓ	3	5	7	11	13	43867
N	5 if $(\frac{p}{3}) = 1$ 6 if $(\frac{p}{3}) = -1$	3	1 if $(\frac{p}{7}) = 1$ 2 if $(\frac{p}{7}) = -1$	1 if $(\frac{p}{11}) = 1$ 2 if $(\frac{p}{11}) = -1$	1	1
m	386	22	1	1	1	0

Table 3 k = 20

ℓ	3	5	7	11	13	283	617
N	5 if $(\frac{p}{3}) = 1$ 6 if $(\frac{p}{3}) = -1$	2	1 if $(\frac{p}{7}) = 1$ 2 if $(\frac{p}{7}) = -1$	1	1	1	1
m	298	13	2	1	1	0	0

Table 4 k = 22

ℓ	3	5	7	13	17	131	593
N	6 if $(\frac{p}{3}) = 1$ 7 if $(\frac{p}{3}) = -1$	2	2	1	1	1	1
m	18	14	37	1	1	0	0

Table 5 k = 22

ℓ	3	5	7	13	17	131	593
N	6 if $(\frac{p}{3}) = 1$ 7 if $(\frac{p}{3}) = -1$	2	2	1	1	1	1
m	18	14	37	1	1	0	0

Table 6 k = 26

ℓ	3	5	7	11	17	19	657931
N	5 if $(\frac{p}{3}) = 1$ 6 if $(\frac{p}{3}) = -1$	2	1 if $(\frac{p}{7}) = 1$ 2 if $(\frac{p}{7}) = -1$	2	1	1	1
m	340	6	2	1	1	1	0

$$
\begin{aligned}
&p \equiv -1 \mod 5^2 \cdot 7^3 \cdot 11 \cdot 3617, \\
&p \equiv -1, 80, 161, 242, 323, 404, 485, 566, 647 \mod 3^6 \quad \text{and} \\
&p \equiv \textit{a non-square} \mod 31;
\end{aligned}
$$

if $a_p(\Delta_{18}) = 0$, then p satisfies

$$
\begin{aligned}
&p \equiv -1 \mod 3^6 \cdot 5^3 \cdot 43867, \\
&p \equiv -1, 19, 31 \mod 7^2, \\
&p \equiv -1, 40, 94, 112, 118 \mod 11^2 \quad \text{and} \\
&p \equiv -1, 4, 10 \mod 13;
\end{aligned}
$$

if $a_p(\Delta_{20}) = 0$, then p satisfies

$$p \equiv -1 \mod 3^6 \cdot 5^2 \cdot 11 \cdot 13 \cdot 283 \cdot 617 \text{ and}$$
$$p \equiv -1, 19, 31 \mod 7^2;$$

if $a_p(\Delta_{22}) = 0$, then p satisfies

$$p \equiv -1 \mod 5^2 \cdot 7^2 \cdot 13 \cdot 17 \cdot 131 \cdot 593 \text{ and}$$
$$p \equiv -1, 728, 1457 \mod 3^7;$$

and if $a_p(\Delta_{26}) = 0$, then p satisfies

$$p \equiv -1 \mod 3^6 \cdot 5^2 \cdot 11 \cdot 17 \cdot 19,$$
$$p \equiv -1, 157780, 578462, 610260, 627364 \mod 657931 \text{ and}$$
$$p \equiv a\ non\text{-}square \mod 7.$$

As discussed above the modular polynomials computed in [1, 17] allow us to calculate the bounds B_k of p for primes p:

Proposition 5.1 *Let the pair (k, B_k) takes the values as in the following table. Then the coefficients $a_p(\Delta_k)$ is non-vanishing for all primes p with*

$$p < B_k.$$

k	B_k
16	4676103180891449
18	354910014891618749
20	6009100058304936449
22	3488537941594810649
26	65980502403749

Proof □

Acknowledgements The author thanks René Schoof who inspires him to work on this topic. Thanks also go to organizers of the first China-Indo number theory conference who provide me the chance to share this work. The author is supported by NSFC (NO: 11601153).

References

1. J. Bosman, On the computation of Galois representations associated to level one modular forms, http://arxiv.org/pdf/0710.1237v1.pdf
2. J.A. Buchmann, H.W. Lenstra Jr., Approximating rings of integers in number fields. J. Théor. Nombres Bordx. **6**(2), 221–260 (1994)
3. P. Deligne, *Formes modulaires et représentations ℓ-adiques*. Lecture Notes in Mathematics, vol. 179 (1971), pp. 139–172
4. T. Dokchitser, V. Dokchitser, Identifying Frobenius elements in Galois groups. Algebra Number Theory **7**(6), 1325–1352 (2013)
5. S.J. Edixhoven, The weight in Serre's conjectures on modular forms. Invent. Math. **109**(3), 563–594 (1992)
6. S.J. Edixhoven, J.-M. Couveignes, R.S. de Jong, F. Merkl, J.G. Bosman, *Computational Aspects of Modular Forms and Galois Representations*. Annals of Mathematics Studies, vol. 176 (Princeton University Press, Princeton, 2011)
7. C. Khare, J.-P. Wintenberger, Serre's modularity conjecture (I), (II). Invent. Math. **178**(3), 485–586 (2009)
8. D.H. Lehmer, The vanishing of Ramanujan's function $\tau(n)$. Duke Math. J. **10**, 429–433 (1947)
9. K.A. Ribet, *Report on Mod ℓ Representations of* $Gal(\bar{\mathbb{Q}}, \mathbb{Q})$, *Motives (Seattle, WA, 1991)* (American Mathematical Society, Providence, 1994), pp. 639–676
10. K.A. Ribet, W.A. Stein, Lectures on Serre's conjectures, *Arithmetic Algebraic Geometry (Park City, UT, 1999)* (American Mathematical Society, Providence, 2001), pp. 143–232
11. R. Schoof, Elliptic curves over finite fields and the computation of square roots mod p. Math. Comput. **44**, 483–494 (1985)
12. J.-P. Serre, Sur la lacunarité des puissances de η. Glasg. Math. J. **27**, 203–221 (1985)
13. J.-P. Serre, Sur les représentations modulaires de degré 2 de $Gal(\overline{\mathbb{Q}}/\mathbb{Q})$. Duke Math. J. **54**(1), 179–230 (1987)
14. W.A. Stein, *Modular Forms, a Computational Approach*. Graduate Studies in Mathematics, vol. 79 (American Mathematical Society, Providence, 2007)
15. H.P.F. Swinnerton-Dyer, *On ℓ-adic Representations and Congruences for Coefficients of Modular Forms (I)*. Lecture Notes in Mathematics, vol. 350 (1973), pp. 1–55
16. H.P.F. Swinnerton-Dyer, *On ℓ-adic Representations and Congruences for Coefficients of Modular Forms (II)*. Lecture Notes in Mathematics, vol. 601 (1977), pp. 63–90
17. P. Tian, Computations of Galois representations associated to modular forms of level one. Acta Arith. **164**, 399–412 (2014)
18. J.R. Wilton, Congruence properties of Ramanujan's function $\tau(n)$. Proc. Lond. Math. Soc. **31**, 1–10 (1930)
19. J.X. Zeng, L.S. Yin, On the computation of coefficients of modular forms: the reduction modulo p approach, http://arxiv.org/pdf/1211.1124.pdf

Hecke Algebras, New Vectors and New Spaces

Soma Purkait

1 Introduction

We report a joint work [2] with E. M. Baruch in which we characterize the space of new forms for $\Gamma_0(N)$ as a common eigenspace of certain Hecke operators which depend on primes dividing the level N.

We shall study a certain p-adic Hecke algebra of functions on $K = \mathrm{GL}_2(\mathbb{Z}_p)$ and describe it using generators and relations. Casselman [3, 4] showed that there is a unique irreducible representation of K which contains a $K_0(p^n)$ fixed vector but does not contain a $K_0(p^k)$ fixed vector for $k < n$. Such a vector is called a new vector and is unique upto scalar multiplication. Using our p-adic Hecke algebra we shall explicitly find for any positive n the $n + 1$ irreducible representations of K which contain a $K_0(p^n)$ fixed vector including the unique representation that contains the "new vector" of level n.

This p-adic Hecke algebra sits inside the endomorphism algebra of $A_{2k}(N)$, the space of adelic automorphic forms of weight $2k$ and level N which is well known [5] to be isomorphic to the classical space of cusp forms $S_{2k}(\Gamma_0(N))$. We use this isomorphism to translate the p-adic Hecke operators to their classical counterparts and obtain relations amongst them. This will lead us to obtain the characterization results about the new spaces.

We present below one of our characterization results.

Theorem 1 *Let N be a square-free positive number. For any prime $p \mid N$, let $Q_p = p^{1-k}U_pW_p$ and $Q'_p = p^{1-k}W_pU_p$. Then the space of new forms $S_{2k}^{\mathrm{new}}(\Gamma_0(N))$ is the intersection of the -1 eigenspaces of Q_p and Q'_p as p varies over the prime divisors of N.*

S. Purkait (✉)
Department of Mathematics, Tokyo Institute of Technology, Tokyo, Japan
e-mail: purkait@math.titech.ac.jp

Z. Liang and C. Aribam (eds.), *The Computational and Theoretical Aspects of Elliptic Curves*, Mathematical Lectures from Peking University,
https://doi.org/10.1007/978-981-13-6664-2_6

For a similar statement for general level N we will later introduce a certain "new" family of Hecke operators.

2 p-adic Hecke Algebras

In this section we will find generators and relations for a Hecke algebra of functions on $K = \mathrm{GL}_2(\mathbb{Z}_p)$ which are bi-invariant with respect to $K_0(p^n)$.

Let G denote the group $\mathrm{GL}_2(\mathbb{Q}_p)$. Let $K_0(p^n)$ be the subgroup of K defined by

$$K_0(p^n) = \left\{ \begin{pmatrix} a & b \\ c & d \end{pmatrix} \in K \ : \ c \in p^n\mathbb{Z}_p \right\}.$$

The subgroup $K_0(p)$ denote the usual Iwahori subgroup.

The Hecke algebra corresponding to $K_0(p^n)$ is defined as:

$$H(G//K_0(p^n)) = \{f \in C_c^\infty(G) : f(kgk') = f(g) \text{ for } g \in G, \ k, \ k' \in K_0(p^n)\},$$

it forms a $\mathbb{C}$-algebra under convolution which, for any $f_1, f_2 \in C_c^\infty(G)$, is defined by

$$f_1 * f_2(h) = \int_G f_1(g) f_2(g^{-1}h) dg = \int_G f_1(hg) f_2(g^{-1}) dg,$$

where dg is the Haar measure on G such that the measure of $K_0(p^n)$ is one.

Let X_g be the characteristic function of the double coset $K_0(p^n)gK_0(p^n)$. Then $H(G//K_0(p^n))$ as a $\mathbb{C}$-vector space is spanned by X_g as g varies over the double coset representatives of G modulo $K_0(p^n)$.

Let $\mu(g)$ denote the number of disjoint left (right) $K_0(p^n)$ cosets in the double coset $K_0(p^n)gK_0(p^n)$. Then the following lemmas are well known [6, Corollary 1.1].

Lemma 2.1 *If $\mu(g)\mu(h) = \mu(gh)$ then $X_g * X_h = X_{gh}$.*

Lemma 2.2 *Let $f_1, f_2 \in H(G//K_0(p^n))$ such that f_1 is supported on $K_0(p^n)x K_0(p^n) = \bigcup_{i=1}^m \alpha_i K_0(p^n)$ and f_2 is supported on $K_0(p^n)yK_0(p^n) = \bigcup_{j=1}^n \beta_j K_0(p^n)$. Then*

$$f_1 * f_2(h) = \sum_{i=1}^m f_1(\alpha_i) f_2(\alpha_i^{-1}h)$$

where the nonzero summands are precisely for those i for which there exist a j such that $h \in \alpha_i \beta_j K_0(p^n)$.

For $t \in \mathbb{Q}_p$ we shall consider the following elements:

$$x(t) = \begin{pmatrix} 1 & t \\ 0 & 1 \end{pmatrix}, \; y(t) = \begin{pmatrix} 1 & 0 \\ t & 1 \end{pmatrix}, \; w(t) = \begin{pmatrix} 0 & -1 \\ t & 0 \end{pmatrix},$$

$$d(t) = \begin{pmatrix} t & 0 \\ 0 & 1 \end{pmatrix}, \; z(t) = \begin{pmatrix} t & 0 \\ 0 & t \end{pmatrix}.$$

Let $N = \{x(t) : t \in \mathbb{Q}_p\}$, $\bar{N} = \{y(t) : t \in \mathbb{Q}_p\}$ and A be the group of diagonal matrices of G. Let $Z_G = \{z(t) : t \in \mathbb{Q}_p^*\}$ denote the center of G.

2.1 The Iwahori Hecke Algebra

We first look at the case when $n = 1$. We have following well-known lemma.

Lemma 2.3 *A complete set of representatives for the double cosets of G mod $K_0(p)$ are given by $d(p^n)z(m)$, $w(p^n)z(m)$ where n, m varies over integers.*

Using triangular decomposition of $K_0(p)$ we obtain following decomposition.

Lemma 2.4 *(1) For $n \geq 0$ we have*

$$K_0(p)d(p^n)K_0(p) = \bigsqcup_{s \in \mathbb{Z}_p/p^n\mathbb{Z}_p} x(s)d(p^n)K_0(p).$$

(2) For $n \geq 1$ we have

$$K_0(p)d(p^{-n})K_0(p) = \bigsqcup_{s \in \mathbb{Z}_p/p^n\mathbb{Z}_p} y(ps)d(p^{-n})K_0(p).$$

(3) For $n \geq 1$ we have

$$K_0(p)w(p^n)K_0(p) = \bigsqcup_{s \in \mathbb{Z}_p/p^{n-1}\mathbb{Z}_p} y(ps)w(p^n)K_0(p).$$

(4) For $n \geq 0$ we have

$$K_0(p)w(p^{-n})K_0(p) = \bigsqcup_{s \in \mathbb{Z}_p/p^{n+1}\mathbb{Z}_p} x(s)w(p^{-n})K_0(p).$$

Let $\mathcal{T}_n = X_{d(p^n)}$, $\mathcal{U}_n = X_{w(p^n)}$ and $\mathcal{Z} = X_{z(p)}$ be elements of the Hecke algebra $H(G//K_0(p))$. Note that $\mathcal{Z}$ commutes with every $f \in H(G//K_0(p))$ and that $\mathcal{Z}^n = X_{z(p^n)}$. We use Lemmas 2.1, 2.2, 2.4 to obtain the following relations in $H(G//K_0(p))$.

Lemma 2.5 *(1) If $n, m \geq 0$ or $n, m \leq 0$, then $\mathcal{T}_n * \mathcal{T}_m = \mathcal{T}_{n+m}$.*
(2) If $n \geq 0$ then $\mathcal{U}_1 * \mathcal{T}_n = \mathcal{U}_{n+1}$ and $\mathcal{T}_n * \mathcal{U}_1 = \mathcal{Z}^n * \mathcal{U}_{1-n}$.
(3) If $n \geq 0$ then $\mathcal{U}_1 * \mathcal{T}_{-n} = \mathcal{U}_{1-n}$ and $\mathcal{T}_{-n} * \mathcal{U}_1 = \mathcal{Z}^{-n} * \mathcal{U}_{1+n}$.
(4) If $n \geq 0$ then $\mathcal{U}_0 * \mathcal{T}_{-n} = \mathcal{U}_{-n}$ and $\mathcal{T}_n * \mathcal{U}_0 = \mathcal{Z}^n * \mathcal{U}_{-n}$.
(5) For $n \in \mathbb{Z}$, $\mathcal{U}_1 * \mathcal{U}_n = \mathcal{Z} * \mathcal{T}_{n-1}$ and $\mathcal{U}_n * \mathcal{U}_1 = \mathcal{Z}^n * \mathcal{T}_{1-n}$.
(6) For $n \geq 1$, $\mathcal{U}_0 * \mathcal{U}_n = \mathcal{T}_n$ and $\mathcal{U}_n * \mathcal{U}_0 = \mathcal{Z}^n * \mathcal{T}_{-n}$.
(7) $\mathcal{U}_0 * \mathcal{U}_0 = (p-1)\mathcal{U}_0 + p$.

Thus we have the following well known theorem.

Theorem 2 *The Iwahori Hecke Algebra $H(G//K_0(p))$ is generated by $\mathcal{U}_0$, $\mathcal{U}_1$ and $\mathcal{Z}$ with the relations:*

1) $\mathcal{U}_1^2 = \mathcal{Z}$,
2) $(\mathcal{U}_0 - p)(\mathcal{U}_0 + 1) = 0$,
3) $\mathcal{Z}$ commutes with $\mathcal{U}_0$ and $\mathcal{U}_1$.

2.2 A Subalgebra

We shall now consider the case $n \geq 2$.

Let $H(K//K_0(p^n))$ denote the subalgebra of $H(G//K_0(p^n))$ consisting of functions supported on K. We shall obtain generators and relations for $H(K//K_0(p^n))$.

We first note the following lemma [4, Lemma 1].

Lemma 2.6 *A complete set of representatives for the double cosets of K mod $K_0(p^n)$ are given by $1,\ w(1),\ y(p),\ y(p^2),\ \ldots\ y(p^{n-1})$.*

Let $\mathcal{U}_0 = X_{w(1)}$ and $\mathcal{V}_r = X_{y(p^r)}$ for $1 \leq r \leq n-1$ be the elements of $H(G//K_0(p^n))$. Then by the above lemma, $H(K//K_0(p^n))$ is spanned by 1, $\mathcal{U}_0$ and $\mathcal{V}_r$ where $1 \leq r \leq n-1$.

We prove the following lemma.

Lemma 2.7

$$K_0(p^n)y(p^r)K_0(p^n) = \bigsqcup_{s \in \mathbb{Z}_p^*/1+p^{n-r}\mathbb{Z}_p} d(s)y(p^r)K_0(p^n).$$

As a consequence of the above lemma and Lemma 2.2 we obtain the following.

Proposition 2.8 *We have the following relations in $H(K//K_0(p^n))$:*

(1) $\mathcal{V}_r^2 = p^{n-r-1}(p-1)(I + \sum_{j=r+1}^{n-1} \mathcal{V}_j) + p^{n-r-1}(p-2)\mathcal{V}_r$.
(2) $\mathcal{V}_r * \mathcal{V}_j = (p-1)p^{n-j-1}\mathcal{V}_r = \mathcal{V}_j * \mathcal{V}_r$ for $r+1 \leq j \leq n-1$.
(3) Let $\mathcal{Y}_{r+1} = I + \sum_{j=r+1}^{n-1} \mathcal{V}_j$. Then

$$\mathcal{V}_r * \mathcal{Y}_{r+1} = p^{n-r-1}\mathcal{V}_r = \mathcal{Y}_{r+1} * \mathcal{V}_r,$$

and so,

$$(\mathcal{V}_r - p^{n-r-1}(p-1))(\mathcal{V}_r + \mathcal{Y}_{r+1}) = 0.$$

For $1 \leq r \leq n-1$, let $\mathcal{Y}_r$ be as before, i.e. $\mathcal{Y}_r = I + \sum_{j=r}^{n-1} \mathcal{V}_j$, take $\mathcal{Y}_n = I$. We have following easy corollary.

Corollary 2.9 *(1)* $\mathcal{Y}_{n-r}^2 = p^r \mathcal{Y}_{n-r}$ *for all* $0 \leq r \leq n-1$.
(2) $\mathcal{Y}_r * \mathcal{Y}_l = p^{n-r}\mathcal{Y}_l = \mathcal{Y}_l * \mathcal{Y}_r$ *for* $r \geq l$.

Next we obtain relations in $H(K//K_0(p^n))$ that involve $\mathcal{U}_0$.

Proposition 2.10 *(1)* $\mathcal{U}_0 * \mathcal{U}_0 = p^{n-1}(p-1)\mathcal{U}_0 + p^n\mathcal{Y}_1$.
(2) $\mathcal{U}_0 * \mathcal{Y}_r = p^{n-r}\mathcal{U}_0 = \mathcal{Y}_r * \mathcal{U}_0$ *for all* $1 \leq r \leq n$.
(3) $\mathcal{U}_0 * (\mathcal{U}_0 - p^n) * (\mathcal{U}_0 + p^{n-1}) = 0$.

Thus we have the following theorem.

Theorem 3 *The algebra* $H(K//K_0(p^n))$ *is an* $n+1$ *dimensional commutative algebra with generators* $\{\mathcal{U}_0, \mathcal{Y}_1, \mathcal{Y}_2, \ldots, \mathcal{Y}_n\}$ *and relations given by Corollary 2.9 and Proposition 2.10.*

We should point out that we have not yet found an analogue of Theorem 2 for $H(G//K_0(p^n))$ for $n \geq 2$. However we would need the following relation later. Let $\mathcal{T}_m = X_{d(p^m)}, \mathcal{U}_m = X_{w(p^m)}, \mathcal{Z} = X_{z(p)}$ be the elements in $H(G//K_0(p^n))$. Then

Lemma 2.11 $(\mathcal{T}_1)^m * \mathcal{U}_m = \mathcal{T}_m * \mathcal{U}_m = \mathcal{Z}^m * \mathcal{U}_0$ *for all* $m \leq n$.

2.3 Representations of K Having a $K_0(p^n)$ Fixed Vector

Let

$$I(n) := Ind_{K_0(p^n)}^K 1 = \{\phi : K \to \mathbb{C} : \phi(k_0 k) = \phi(k) \text{ for } k_0 \in K_0(p^n),\ k \in K\}.$$

Then $I(n)$ is a right representation of K, via right translation, denoted by π_R, where $\pi_R(k)(\phi)(k') = \phi(k'k)$, and the dimension of this representation is $[K : K_0(p^n)] = p^{n-1}(p+1)$. It follows from Frobenius Reciprocity that every (smooth) irreducible representation of K which has a nonzero $K_0(p^n)$ fixed vector is isomorphic to a subrepresentation of $I(n)$. We shall therefore decompose $I(n)$ into sum of irreducible representations.

Lemma 2.12 *We have* $I(n)^{K_0(p^n)} = H(K//K_0(p^n))$ *and consequently the dimension of* $I(n)^{K_0(p^n)}$ *is* $n+1$.

Further, using induction argument and Frobenius reciprocity we can check that the representation $I(n)$ is a sum of $n+1$ distinct irreducible representations.

We shall now explicitly describe the irreducible subrepresentations of $I(n)$. Let us consider the action π_L of $H(K//K_0(p^n))$ on $I(n)$: for $f \in H(K//K_0(p^n))$ and $\phi \in I(n)$ set

$$\pi_L(f)(\phi)(g) = \int_K f(k)\phi(k^{-1}g)dk \quad \text{for all } g \in K.$$

In particular, if $\phi \in I(n)^{K_0(p^n)}$ which by Lemma 2.12 is same as the algebra $H(K//K_0(p^n))$ then we have $\pi_L(f)(\phi) = f * \phi$. It is easy to check that the action π_L commutes with the action π_R. It now follows by Schur's Lemma that for each $f \in H(K//K_0(p^n))$ the operator $\pi_L(f)$ acts as a scalar operator on an irreducible subrepresentation of $I(n)$. We shall use this to distinguish the irreducible components of $I(n)$ as follows.

If σ is any irreducible subrepresentation of $I(n)$ then σ contains a $K_0(p^n)$ fixed vector, that is there exists a non-zero vector $v_\sigma \in \sigma \cap I(n)^{K_0(p^n)}$. Thus v_σ is a linear combination of $\mathcal{U}_0$ and $\mathcal{Y}_r$ for $1 \le r \le n$. Since $\pi_L(f)$ acts as a scalar for every $f \in H(K//K_0(p^n))$ the vector v_σ will be an eigenvector under the action of $\pi_L(\mathcal{U}_0)$ and $\pi_L(\mathcal{Y}_r)$ for all $1 \le r \le n$. For each σ we can compute these eigenvectors v_σ and their corresponding eigenvalues using the relations in Corollary 2.9 and Proposition 2.10. Thus we obtain the following proposition.

Proposition 2.13 *A basis of eigenvectors for $H(K//K_0(p^n))$ under the above action is given by:*
$v_1 = \mathcal{U}_0 + \mathcal{Y}_1$,
$v_2 = \mathcal{U}_0 - p\mathcal{Y}_1$,
$w_k = \mathcal{Y}_k - p\mathcal{Y}_{k+1}$ *for* $1 \le k \le n-1$,
with eigenvalues given as follows:
$\mathcal{U}_0 * v_1 = p^n v_1$, $\mathcal{Y}_i * v_1 = p^{n-i} v_1$ *for all* $1 \le i \le n$,
$\mathcal{U}_0 * v_2 = -p^{n-1} v_2$, $\mathcal{Y}_i * v_2 = p^{n-i} v_2$ *for all* $1 \le i \le n$,
$\mathcal{U}_0 * w_k = 0$, $\mathcal{Y}_i * w_k = 0$ *for all* $1 \le i \le k$,
$\mathcal{Y}_i * w_k = p^{n-i} w_k$ *for all* $k < i \le n$.

Corollary 2.14 *The representation $I(n)$ is a sum of $n+1$ irreducible subspaces given by:* $S_1 = \mathrm{Span}(\pi_R(K)v_1)$, $S_2 = \mathrm{Span}(\pi_R(K)v_2)$ *and* $T_k = \mathrm{Span}(\pi_R(K)w_k)$ *where* $1 \le k \le n-1$ *such that* $\dim(S_1) = 1$, $\dim(S_2) = p$, $\dim(T_k) = p^{k-1}(p^2 - 1)$. *Consequently, T_{n-1} is the unique irreducible representation of K that has a $K_0(p^n)$ fixed vector w_{n-1} but does not have $K_0(p^k)$ fixed vector for $k < n$ that is, w_{n-1} is the "new" vector of level n.*

3 Translation from the Adelic Setting to the Classical Setting

Let $G_\infty = \mathrm{GL}_2(\mathbb{R})^+$. Then G_∞ acts on the upper half plane $\mathbb{H}$ in a standard way. For $g = \begin{pmatrix} a & b \\ c & d \end{pmatrix} \in G_\infty$, $z \in \mathbb{H}$ and functions f on $\mathbb{H}$ recall the automorphy factor and the slash operator $|_{2k}g$,

$$j(g, z) = det(g)^{-1/2}(cz + d), \quad f|_{2k}g = j(g, z)^{-2k} f\left(\frac{az + b}{cz + d}\right).$$

Let N be a positive integer and $K_p = K_0(p^\alpha)$ for a prime p such that $p^\alpha \| N$. Let K_f be the subgroup of $\mathrm{GL}_2(\mathbb{A})$ defined by

$$K_f(N) = \prod_{q<\infty} K_q.$$

By the strong approximation theorem we have

$$\mathrm{GL}_2(\mathbb{A}) = \mathrm{GL}_2(\mathbb{Q}) G_\infty K_f(N).$$

Denote by $A_{2k}(N)$ the space of functions $\Phi \in L^2(Z_\mathbb{A}\,\mathrm{GL}_2(\mathbb{Q})\backslash \mathrm{GL}_2(\mathbb{A}))$ satisfying the following properties:

(1) $\Phi(gk) = \Phi(g)$ for all $g \in \mathrm{GL}_2(\mathbb{A}),\ k \in K_f(N)$.

(2) $\Phi(gr(\theta)) = e^{-i2k\theta}\Phi(g)$ where $r(\theta) = \begin{pmatrix} cos\theta & -sin\theta \\ sin\theta & cos\theta \end{pmatrix} \in \mathrm{SO}(2)$.

(3) Φ is smooth as a function of G_∞ and satisfies the differential equation $\Delta\Phi = -k(k-1)\Phi$ where Δ is the Casimir operator.

(4) Φ is cuspidal, that is $\int_{\mathbb{Q}\backslash\mathbb{A}} \Phi\left(\begin{pmatrix} 1 & a \\ 0 & 1 \end{pmatrix} g\right) da = 0$ for all $g \in \mathrm{GL}_2(\mathbb{A})$.

By Gelbart [5, Proposition 3.1] there exists an isomorphism

$$A_{2k}(N) \to S_{2k}(\Gamma_0(N))$$

given by $\Phi \mapsto f_\Phi$ where for $z \in \mathbb{H}$,

$$f_\Phi(z) = \Phi(g_\infty) j(g_\infty, i)^{2k}$$

where $g_\infty \in G_\infty$ is such that $g_\infty(i) = z$. The inverse map is given by $f \mapsto \Phi_f$ where for $g \in \mathrm{GL}_2(\mathbb{A})$ if $g = \gamma g_\infty k$ (using strong approximation),

$$\Phi_f(g) = f(g_\infty(i)) j(g_\infty, i)^{-2k}.$$

This isomorphism induces a ring isomorphism of spaces of linear operators by

$$q : \mathrm{End}_{\mathbb{C}}(A_{2k}(N)) \rightarrow \mathrm{End}_{\mathbb{C}}(S_{2k}(\Gamma_0(N)))$$

given by

$$q(\mathcal{T})(f) = f_{\mathcal{T}(\Phi_f)}.$$

Let $N = p^n M$ where p is a prime coprime to M and $G = \mathrm{GL}_2(\mathbb{Q}_p)$. We note that the $H(G//K_0(p^n))$ is a subalgebra of $\mathrm{End}_{\mathbb{C}}(A_{2k}(N))$ via the following action:

$$\text{for } \mathcal{T} \in H(G//K_0(p^n)) \text{ and } \Phi \in A_{2k}(N), \quad \mathcal{T}(\Phi)(g) = \int_G \mathcal{T}(x)\Phi(gx)dx.$$

We have following proposition.

Proposition 3.1 *Let $N = p^n M$ where $n \geq 1$ and $p \nmid M$. Let $f \in S_{2k}(\Gamma_0(N))$. Consider operators $\mathcal{T}_1, \mathcal{U}_m \in H(G//K_0(p^n))$ where $m \leq n$. If $n \geq 2$, further consider $\mathcal{V}_r \in H(G//K_0(p^n))$ where $1 \leq r \leq n-1$. Then,*

(1) $q(\mathcal{T}_1)(f)(z) = p^{-k} \sum_{s=0}^{p-1} f((z+s)/p) = \tilde{U}_p(f)(z)$.
(2) If $f \in S_{2k}(\Gamma_0(p^r M))$ where $r \leq n$ then $q(\mathcal{U}_r)(f)(z) = p^{n-r} f|_{2k} W_{p^r}(z)$ where $W_{p^r} = \begin{pmatrix} p^r\beta & 1 \\ p^r M\gamma & p^r \end{pmatrix}$ is an integer matrix of determinant p^r. In particular, $q(\mathcal{U}_n)(f)(z) = f|_{2k} W_{p^n}(z)$.
(3) $q(\mathcal{V}_r)(f)(z) = \sum_{s \in \mathbb{Z}_p^*/1+p^{n-r}\mathbb{Z}_p} f|_{2k} A_s$ where $A_s \in \mathrm{SL}_2(\mathbb{Z})$ is any matrix of the form $\begin{pmatrix} a_s & b_s \\ p^r M & p^{n-r} - sM \end{pmatrix}$.
(4) If $f \in S_{2k}(\Gamma_0(p^r M))$ then $q(\mathcal{V}_r)(f) = p^{n-r-1}(p-1)f$, consequently, $q(\mathcal{Y}_r)(f) = p^{n-r} f$.

Remark 1 The operator $q(\mathcal{U}_n)$ is the usual Atkin-Lehner operator W_{p^n} while the operator $q(\mathcal{T}_1)$ is the operator $\tilde{U}_p = p^{1-k} U_p$ where U_p is the usual Hecke operator. It is obvious that $q(\mathcal{Z})$ is the identity operator.

Let $N = pM$ where $p \nmid M$. Let $Q_p = q(\mathcal{U}_0)$ where $\mathcal{U}_0 \in H(G//K_0(p))$. Then using Lemma 2.5 we have

Corollary 3.2 *$Q_p = p^{1-k} U_p W_p$ and $(Q_p - p)(Q_p + 1) = 0$.*

Now consider $N = p^n M$ where $n \geq 2$. Let $Q_{p^m} = (\tilde{U}_p)^m W_{p^m}$ for $m \leq n$ where W_{p^m} is the Atkin-Lehner operator on $S_{2k}(\Gamma_0(p^m M))$. Using Lemma 2.11 and Propositions 3.1 and 2.10 we have

Corollary 3.3 *For $\mathcal{U}_0 \in H(G//K_0(p^n))$, we have $Q_{p^n} = q(\mathcal{U}_0)$ and hence $Q_{p^n}(Q_{p^n} - p^n)(Q_{p^n} + p^{n-1}) = 0$. Further for $m \leq n$ we have $Q_{p^n} = (\tilde{U}_p)^m q(\mathcal{U}_m)$, hence if $f \in S_{2k}(\Gamma_0(p^m M)) \subseteq S_{2k}(\Gamma_0(N))$ then $Q_{p^n}(f) = p^{n-m} Q_{p^m}(f)$.*

Let $S_{p^n,r} = q(\mathcal{Y}_r)$ where $\mathcal{Y}_r \in H(G//K_0(p^n))$, $1 \le r \le n$. Using relations in Corollary 2.9, we have

Corollary 3.4 $S_{p^n,r}(S_{p^n,r} - p^{n-r}) = 0$ *for* $1 \le r \le n$.

Let $Q'_{p^n} = W_{p^n} Q_{p^n} W_{p^n}^{-1}$ and $S'_{p^n,r} = W_{p^n} S_{p^n,r} W_{p^n}^{-1}$. Then Q'_{p^n} and $S'_{p^n,r}$ also satisfy the above cubic and quadratic relations.

4 Main Results

The following is a restatement (slightly general) of Theorem 1.

Theorem 1 *Let* $N = M_1^2 M$ *where* M_1 *and* M *are square free and coprime. Then* $f \in S_{2k}^{\text{new}}(\Gamma_0(N))$ *if and only if* $Q_p(f) = -f = Q'_p(f)$ *for all primes p dividing M and* $Q_{p^2}(f) = 0 = Q'_{p^2}(f)$ *for all primes p dividing* M_1.

For a general N we need to use the family of operators $S_{p^n,r}$ to obtain a similar characterization result.

Theorem 4 *Let N be a positive integer. Then the space of new forms* $S_{2k}^{\text{new}}(\Gamma_0(N))$ *is the intersection of the* -1 *eigenspaces of* Q_p *and* Q'_p *where p varies over the primes such that* $p \| N$ *and the* 0 *eigenspaces of* $S_{p^\gamma,\gamma-1}$ *and* $S'_{p^\gamma,\gamma-1}$ *for primes p such that* $p^\gamma \| N$ *with* $\gamma \ge 2$. *That is,* $f \in S_{2k}^{\text{new}}(\Gamma_0(N))$ *if and only if* $Q_p(f) = -f = Q'_p(f)$ *for all primes p such that* $p \| N$ *and* $S_{p^\gamma,\gamma-1}(f) = 0 = S'_{p^\gamma,\gamma-1}(f)$ *for all primes p such that* $p^\gamma \| N$ *for* $\gamma \ge 2$.

Let $q = e^{2\pi i z}$ and $f(z) = \sum_{n=1}^{\infty} a_n q^n \in S_{2k}(\Gamma_0(m))$. Let p be an odd prime. Define

$$R_p(f)(z) = \sum_{n=1}^{\infty} \left(\frac{n}{p}\right) a_n q^n, \qquad R_\chi(f)(z) = \sum_{n=1}^{\infty} \left(\frac{-1}{n}\right) a_n q^n.$$

By [1, Lemma 33], R_p and R_χ are operators on $S_{2k}(\Gamma_0(m))$ provided that $p^2 \mid m$ and $16 \mid m$ respectively.

Theorem 5 *Let* $N = 2^\beta M_1 M_2$ *where* $M_1 M_2$ *is odd such that* M_1 *is square free and any prime divisor of* M_2 *divides it with a power at least* 2. *Let* $\beta \ge 4$. *Then* $f \in S_{2k}^{\text{new}}(\Gamma_0(N))$ *if and only if* $Q_p(f) = -f = Q'_p(f)$ *for all primes p dividing* M_1, $(R_\chi)^2(f) = f$ *and* $(R_p)^2(f) = f$ *for all primes p dividing* M_2, *and* $S_{p^\gamma,\gamma-1}(f) = 0$ *for all primes p such that* $p^\gamma \| 2^\beta M_2$.

4.1 Sketch of Proof

We shall now sketch a proof of the Theorem 4 in the case when a prime divisor of N divides it with a power at least 2. Let $N = p^n M$ where $n \geq 2$ and $(p, M) = 1$. Recall the family of operators that we defined: for $1 \leq r \leq n$,

$$S_{p^n,r}(f) = f + \sum_{j=r}^{n-1} \sum_{s \in \mathbb{Z}_p^*/1+p^{n-j}\mathbb{Z}_p} f|_{2k} A_{s,j},$$

where $A_{s,j} = \begin{pmatrix} a_{s,j} & b_{s,j} \\ p^j M & p^{n-j} - sM \end{pmatrix} \in \mathrm{SL}_2(\mathbb{Z})$, and

$$S_{p^n,r}(S_{p^n,r} - p^{n-r}) = 0.$$

We have the following lemma.

Lemma 4.1 *For $1 \leq r \leq n$, a set of right coset representatives for $\Gamma_0(N)$ in $\Gamma_0(p^r M)$ consists of the identity element and elements of the form*

$$A_{s,j} = \begin{pmatrix} a_{s,j} & b_{s,j} \\ p^j M & p^{n-j} - sM \end{pmatrix} \text{ where } r \leq j \leq n-1 \text{ and } s \in \mathbb{Z}_p^*/1 + p^{n-j}\mathbb{Z}_p.$$

Consequently, the operator $S_{p^n,r}$ takes the space $S_{2k}(\Gamma_0(N))$ to $S_{2k}(\Gamma_0(p^r M))$.

Thus we have the following corollary.

Corollary 4.2 *For $1 \leq r \leq n$, the p^{n-r} eigenspace of $S_{p^n,r}$ is precisely the subspace $S_{2k}(\Gamma_0(p^r M))$.*

Proposition 4.3 *Let $1 \leq r \leq n$. Then for each $r < \alpha \leq n$, the space $S_{2k}^{\text{new}}(\Gamma_0(p^\alpha M))$ is contained in the 0 eigenspace of $S_{p^n,r}$.*

Proof For a prime q with $(q, N) = 1$, the Hecke operator T_q on $S_{2k}(\Gamma_0(N))$ corresponds to the characteristic function of $\mathrm{GL}_2(\mathbb{Z}_q) \begin{pmatrix} q & 0 \\ 0 & 1 \end{pmatrix} \mathrm{GL}_2(\mathbb{Z}_q)$ which belongs to the q-adic Hecke algebra. Since $\mathcal{Y}_r$ belongs to the p-adic Hecke algebra $H(K_0(p^n))$, it follows that the operators $S_{p^n,r}$ and T_q on $S_{2k}(\Gamma_0(N))$ commute.

Let $r < \alpha \leq n$ and $f \in S_{2k}^{\text{new}}(\Gamma_0(p^\alpha M))$ be a primitive form. Thus f is an eigenform with respect to T_q for any q coprime to N. Now since $S_{p^n,r}$ and T_q commute we get that $S_{p^n,r}(f)$ is also an eigenfunction with respect to all such T_q having the same eigenvalue as f.

By the above lemma, $S_{p^n,r}(f) \in S_{2k}(\Gamma_0(p^r M))$ and as $r < \alpha$, it is an old form in the space $S_{2k}(\Gamma_0(p^\alpha M))$. It now follows from Atkin and Lehner that $S_{p^n,r}(f) = 0$.

Recall the shift operator $V(d)$ which takes a cusp form $f(z)$ of level n to cusp form $f(dz)$ of level nd. We have the following lemma.

Lemma 4.4 *For $1 \leq r \leq n$, the operator W_{p^n} maps $S_{2k}(\Gamma_0(p^r M))$ onto $V(p^{n-r})S_{2k}(\Gamma_0(p^r M))$.*

Consequently for $1 \leq r \leq n$, the p^{n-r} eigenspace of $S'_{p^n,r}$ is precisely the space $V(p^{n-r})S_{2k}(\Gamma_0(p^r M))$.

Applying above results to the case $r = n - 1$ we have the following corollary.

Corollary 4.5 *The space $S_{2k}(\Gamma_0(p^{n-1}M))$ is the p eigenspace of $S_{p^n,n-1}$ and $V(p)S_{2k}(\Gamma_0(p^{n-1}M))$ is the p eigenspace of $S'_{p^n,n-1}$. Moreover, the space $S_{2k}^{\text{new}}(\Gamma_0(N))$ is contained in the intersection of the 0 eigenspaces of $S_{p^n,n-1}$ and $S'_{p^n,n-1}$.*

Next we have the following proposition.

Proposition 4.6 *The operators $S_{p^n,n-1}$ and $S'_{p^n,n-1}$ are self-adjoint with respect to Petersson inner product.*

Now we give a proof of Theorem 4 where $N = q_1^{\alpha_1} q_2^{\alpha_2} \cdots q_s^{\alpha_s}$ withq_j distinct primes and $\alpha_j \geq 2$ for all $1 \leq j \leq s$. We have already seen one side implication. Conversely suppose $f \in S_{2k}(\Gamma_0(N))$ is such that $S_{q_j^{\alpha_j},\alpha_j-1}(f) = 0 = S'_{q_j^{\alpha_j},\alpha_j-1}(f)$ for all $1 \leq j \leq s$. It follows from from Corollary 4.5 that for each $1 \leq j \leq s$, the subspace $S_{2k}(\Gamma_0(N/q_j))$ is contained in the q_j eigenspace of $S_{q_j^{\alpha_j},\alpha_j-1}$ and $V(q_j)S_{2k}(\Gamma_0(N/q_j))$ is contained in the q_j eigenspace of $S'_{q_j^{\alpha_j},\alpha_j-1}$.

Since $S_{q_j^{\alpha_j},\alpha_j-1}$, $S'_{q_j^{\alpha_j},\alpha_j-1}$ are self-adjoint operators we get that f is orthogonal to $S_{2k}(\Gamma_0(N/q_j))$ and $V(q_j)S_{2k}(\Gamma_0(N/q_j))$ for each prime divisor q_j of N. Thus f is orthogonal to the old space, that is, $f \in S_{2k}^{\text{new}}(\Gamma_0(N))$.

Acknowledgements It was a pleasure to attend the first China-Indo conference on Elliptic curves in the BICMR where I presented these results. I am grateful for the invitation.

References

1. A.O.L. Atkin, J. Lehner, Hecke operators on $\Gamma_0(m)$. Math. Ann. **185**, 134–160 (1970)
2. E.M. Baruch, S. Purkait, Hecke algebras, new vectors and new forms on $\Gamma_0(m)$. Math. Zeit. (2017), https://doi.org/10.1007/s00209-017-1842-y
3. W. Casselman, On some results of Atkin and Lehner. Math. Ann. **201**, 301–314 (1973)
4. W. Casselman, The restriction of a representation of $\mathrm{GL}_2(k)$ to $\mathrm{GL}_2(\mathfrak{o})$. Math. Ann. **206**, 311–318 (1973)
5. S. Gelbart, *Automorphic forms on adele groups*. Annals of Mathematics Studies, vol. 83 (Princeton University Press, Princeton, 1975)
6. R. Howe, Affine-like Hecke algebras and p-adic representation theory, in *Iwahori-Hecke Algebras and their Representation Theory*. Lecture Notes in Mathematics, vol. 1804 (2002), pp. 27–69

A Note on a Formula of Special Values of Dirichlet L-Functions

Gongrong Yang

Abstract In this paper, we get a new formula on special values of Dirichlet L-function at non-positive integers by generalized Euler polynomials. This formula can be simplified for some special conductors. As an example, the resulting formulas are equivalent to Shimura's formulas when $p = 2, 3$.

1 Introduction

In number theory, the special values of L-functions has been studied for a long time. This topic has caused extensive concern in this field, for its close connection to famous problems such like the Riemann hypothesis, distribution of prime numbers.

In the classical case, the values of zeta functions at positive integer points are always expressed by Bernoulli numbers or Bernoulli polynomials. Accurately, let χ be a nontrivial primitive Dirichlet character of the conductor $d > 2$, k a positive integer such that $\chi(-1) = (-1)^k$. The Dirichlet L-function attached to χ is defined by

$$L(s, \chi) = \sum_{n=1}^{\infty} \chi(n) n^{-s}.$$

Let the notation $\mathbf{e}(z) = e^{2\pi i z} (z \in \mathbb{C})$, the Gauss sum $G(\chi)$ of χ

$$G(\chi) = \sum_{a=1}^{d} \chi(a) \mathbf{e}(a/d).$$

There is a well known formula

$$2k!(2\pi i)^{-k} G(\overline{\chi}) L(k, \chi) = -\sum_{a=1}^{q} \overline{\chi}(a) B_k(a/d). \tag{1.1}$$

G. Yang (✉)
Beijing International Center for Mathematical Research, Peking University, Beijing, China
e-mail: yanggr@rdfz.cn

Z. Liang and C. Aribam (eds.), *The Computational and Theoretical Aspects of Elliptic Curves*, Mathematical Lectures from Peking University,
https://doi.org/10.1007/978-981-13-6664-2_7

Here, $B_k(a/d)$ is the kth Bernoulli polynomial which is defined by

$$\frac{ze^{tz}}{e^z-1}=\sum_{n=0}^{\infty}\frac{B_n(t)}{n!}z^n. \tag{1.2}$$

Recently, Shimura [1–5] considered the expression of Dirichlet L-functions again. He used Euler polynomials instead of Bernoulli polynomials to calculate the special values of Dirichlet L-functions. Precisely, Shimura [4] shows a new formula for $L(k,\chi)$:

$$(k-1)!(2\pi i)^{-k}G(\overline{\chi})L(k,\chi)=\frac{1}{2(2^k-\chi(2))}\sum_{a=1}^{q}\overline{\chi}(a)E_{1,k-1}(2a/d) \tag{1.3}$$

where $q=[(d-1)/2]$. $E_{c,k-1}(2a/d)$ is the $(k-1)$th generalized Euler polynomial which is defined by

$$\frac{(1+c)e^{tz}}{e^z+c}=\sum_{n=0}^{\infty}\frac{E_{c,n}(t)}{n!}z^n. \tag{1.4}$$

Comparing (1.1) and (1.4), Shimura believes (1.4) is better, because from the computational viewpoint, $E_{1,k-1}(t)$ is a polynomial of degree $k-1$, whereas $B_k(t)$ is of degree k. So he tried to get more applications of L-functions by Euler polynomials, such as the general Hurwitz zeta function [1] Th0.2, the Dirichlet L-functions for special conductor [1] Th0.3, [5] Th1.4, the class number h_K for an imaginary quadratic field K [4] (5.9), and so on.

This is a new way to study L-functions. There are also many new results using this method, such as Michael J. Dancs, Tian-Xiao He [6],Yoshinori Hamahata [7], Takeyun Kim [8–10], Qiuming Luo [11].

In this paper, inspired by Shimura's work, we define the general Dirichlet L-function for the character χ and parameter γ by

$$L(s,\chi,\gamma)=\sum_{n=1}^{\infty}\gamma^n\chi(n)n^{-s}.$$

Then, we will get another expression of general L-function at positive integer points, using cyclotomic units instead of Gauss Sum.

Our main result is that

Theorem 1 *The values of Dirichlet L-function at non-positive integer points can be written as*

$$L(-m,\chi,\gamma)=\frac{d^m}{1+c^{-1}}\sum_{a=1}^{d-1}\gamma^a\chi(a)E_{c,m}\left(\frac{a}{d}\right) \tag{1.5}$$

and

$$L(-m,\chi) = \frac{d^m}{\chi(p)p^{1+m}-1}\sum_{i=1}^{p-1}\frac{1}{1+c_i^{-1}}\sum_{a=1}^{d-1}\mathbf{e}(a\cdot i/p)\chi(a)E_{c_i,m}\left(\frac{a}{d}\right) \quad (1.6)$$

where $c_i = -\mathbf{e}(-d\cdot i/p)$.

The Eq. (1.6) is a natural generalization of Shimura's formula [1] [(0.6)]. Shimura has other formulas for Dirichlet L-function at positive integer points using the Gauss sum. Instead, our formula at non-positive integer points uses cyclotomic units.

In the case $p = 2$, $m = k-1$ and $d = 2q+1$ with $q \in \mathbb{Z}_+$, it becomes

$$L(1-k,\chi) = \frac{d^{k-1}}{2^k\chi(2)-1}\sum_{b=1}^{q}(-1)^b\chi(b)E_{1,k-1}(b/d) \quad (1.7)$$

which appeared in [1, Theorem 0.3] Th0.3 and [4] Th4.14, iii. Shimura got many corollaries for special d such as [4] Co 4.16, Th 6.3, Th 6.6, and so on.

Although there are some connections between Bernoulli types and Euler types, such as

$$B_n = \sum_{k=0}^{n-1}\binom{n-1}{k}\frac{n}{4^n-2^n}E_k, \quad n = 2,4,6,\ldots; \quad (1.8)$$

$$E_{1,n}(t) = t^n - \sum_{k=1}^{[(n+1)/2]}\binom{n}{2k-1}(2^{2k}-1)\frac{B_{2k}}{k}t^{n+1-2k}, \quad n \in \mathbb{N}. \quad (1.9)$$

It is not easy to get our formulas from the formulas expressing Dirichlet L-functions by Bernoulli polynomials.

2 The Formula on L-Functions

Let χ be a nontrivial primitive Dirichlet character of conductor $d > 2$, $s \in \mathbb{C}$ and $\gamma \in \mathbb{C}, 0 < |\gamma| \leq 1$. The general L-function $L(s,\chi,\gamma)$ is defined by

$$L(s,\chi,\gamma) = \sum_{n=1}^{\infty}\chi(n)\gamma^n n^{-s}. \quad (2.1)$$

We have

$$L(s,\chi,\gamma)=\sum_{n=1}^{\infty}\chi(n)\gamma^n n^{-s}=\sum_{a=1}^{d-1}\sum_{q=0}^{\infty}\chi(a+qd)\gamma^{a+qd}(a+qd)^{-s}$$
$$=d^{-s}\sum_{a=1}^{d-1}\chi(a)\gamma^a\sum_{q=0}^{\infty}(\gamma^d)^q(\frac{a}{d}+q)^{-s}$$
$$=d^{-s}\sum_{a=1}^{d-1}\chi(a)\gamma^a\zeta(s,\frac{a}{d},\gamma^d), \tag{2.2}$$

where $\zeta(s,x,\gamma)$ is the general Hurwitz zeta function defined by

$$\zeta(s,x,\gamma)=\sum_{n=0}^{\infty}\gamma^n(x+n)^{-s}. \tag{2.3}$$

It is first introduced by Lerch [12], thoroughly discussed by Hurwitz [13]. Shimura [1] developed this formula with Euler polynomial. From the formula (2.2) and Hurwitz and Shimura's results, we have

Proposition 2 *The general L-function $L(s,\chi,\gamma)$ converges when $Re(s)>1$ and $0<|\gamma|\leq 1$. And it can analytically continue to the whole s-plane with $s=1$ as a pole. Moreover, it is also defined for $\gamma\in\mathbb{C}\backslash(1,\infty)$.*

If $\gamma^d=1$, it will return to the ordinary Hurwitz zeta function and the value of non-positive point can be described by Bernoulli polynomial.

If $\gamma^d\neq 1$, we should use Euler polynomial to describe the value of non-positive point. Shimura [1] Th0.2 shows that for any $k\in Z_+$, $Re(a)>0$, $\gamma\in\mathbb{C}$ and $\gamma\notin\{x\in\mathbb{R}|x\geq 1\}$,

$$\zeta(1-k,a,\gamma)=\frac{E_{-\gamma^{-1},k-1}(a)}{1-\gamma^{-1}}. \tag{2.4}$$

So, we have

$$L(1-k,\chi,\gamma)=\frac{d^{k-1}}{1+c^{-1}}\sum_{a=1}^{d-1}\gamma^a\chi(a)E_{c,k-1}\left(\frac{a}{d}\right), \tag{2.5}$$

where $c=-\gamma^{-d}\neq -1$, which is the formula (1.5).

Let p be a prime number such that $p\nmid d$ and $\omega=\mathbf{e}(1/p)$. It is clear that

$$\sum_{i=0}^{p-1}(\omega^i)^n=\begin{cases}p, & p\mid n\\ 0, & p\nmid n.\end{cases} \tag{2.6}$$

In fact, for any positive integer number p, such that $(p, d) = 1$, the result (2.6) is also held. It turns out that

$$\sum_{i=0}^{p-1} L(s, \chi, \omega^i) = \sum_{i=0}^{p-1}\sum_{n=1}^{\infty} \chi(n)(\omega^i)^n n^{-s} = \sum_{n=1}^{\infty} \chi(n)n^{-s} \sum_{i=0}^{p-1}(\omega^i)^n$$
$$= \sum_{p|n} p\chi(n)n^{-s} = \sum_{m=1}^{\infty} p\chi(pm)(pm)^{-s}$$
$$= \chi(p)p^{1-s}\sum_{m=1}^{\infty}\chi(m)m^{-s} = \chi(p)p^{1-s}L(s, \chi) \tag{2.7}$$

Note that $L(s, \chi, 1) = L(s, \chi)$. We can get

$$L(s, \chi) = \frac{1}{\chi(p)p^{1-s} - 1}\sum_{i=1}^{p-1} L(s, \chi, \omega^i). \tag{2.8}$$

Putting $s = 1 - k$ and using (2.5) to express $L(1 - k, \chi, \omega^i)$, we get

$$L(1 - k, \chi) = \frac{d^{k-1}}{\chi(p)p^k - 1}\sum_{i=1}^{p-1}\frac{1}{1 + c_i^{-1}}\sum_{a=1}^{d-1} \mathbf{e}(a \cdot i/p)\chi(a)E_{c_i,k-1}\left(\frac{a}{d}\right), \tag{2.9}$$

where $c_i = -\mathbf{e}(-d \cdot i/p)$ as we claimed by formula (1.6).

In this proof, the the L-function $L(s, \chi)$ first is converted into the general L-function $L(s, \chi, \gamma)$, then into the general Hurwitz zeta function $\zeta(s, \chi, \gamma)$, and finally into the Euler polynomial $E_{c,m}(t)$. The calculation may be cumbersome, but unambiguous.

3 The Formula for Special Values

However the complex summation may bring difficulties in actual computation. So we should simplify the formula for some selected modulus d according to the symmetry of Euler polynomials and characters. Let $\chi(-1) = (-1)^{\delta(\chi)}$. It is obvious that $\chi(d - a) = (-1)^{\delta(\chi)}\chi(a)$. From [4] (4.3f), we get

$$E_{c,n}(1 - t) = (-1)^n E_{c^{-1},n}(t). \tag{3.1}$$

We give two examples for $p = 2$ or 3.

(i) $d = 2q + 1$. This case has already been considered by Shimura.

In this case, we can choose $p = 2$ and trivially $\omega = \mathbf{e}(1/2) = -1$, $c = -\omega^{-1} = 1$. Then

$$\begin{aligned}
L(-m,\chi) &= \frac{d^m}{2(2^{m+1}\chi(2)-1)}\sum_{a=1}^{d-1}\mathbf{e}(a/2)\chi(a)E_{1,m}(\frac{a}{d}) \\
&= \frac{d^m}{2(2^{m+1}\chi(2)-1)}\sum_{a=1}^{q}\left(\mathbf{e}(a/2)\chi(a)E_{1,m}(\frac{a}{d})+\mathbf{e}((d-a)/2)\chi(d-a)E_{1,m}\left(1-\frac{a}{d}\right)\right) \\
&= \frac{d^m}{2(2^{m+1}\chi(2)-1)}\sum_{a=1}^{q}\left((-1)^a\chi(a)E_{1,m}(\frac{a}{d})+(-1)^{d-a}(-1)^{\delta(\chi)}\chi(a)(-1)^{-m}E_{1,m}\left(\frac{a}{d}\right)\right) \\
&= \frac{[1+(-1)^{\delta(\chi)+m+1}]d^m}{2(2^{m+1}\chi(2)-1)}\sum_{a=1}^{q}(-1)^a\chi(a)E_{1,m}\left(\frac{a}{d}\right)
\end{aligned}$$

It is clear that if $m \equiv \delta(\chi) \bmod 2$, $-m$ will be trivial zero point. If $m \equiv \delta(\chi)-1 \bmod 2$, the formula becomes

$$L(-m,\chi) = \frac{d^m}{2^{m+1}\chi(2)-1}\sum_{a=1}^{q}(-1)^a\chi(a)E_{1,m}\left(\frac{a}{d}\right), \tag{3.2}$$

which is exactly the formula in [1] Th0.3(0.6).

(ii) Suppose d is prime to 3;

In this case, we can choose $p = 3$, $\omega = \mathbf{e}(1/3)$. Put $\tilde{L}(-m,\chi) = \frac{\chi(3)3^{m+1}-1}{d^m}L(-m,\chi)$. Then $c_1 = -\mathbf{e}(-\frac{d}{3}) = -\omega^{-d} = -\omega^{2d} = c_2^{-1}$. We have

$$\begin{aligned}
\tilde{L}(-m,\chi) &= \sum_{a=1}^{d-1}\left(\frac{1}{1+c_1^{-1}}\mathbf{e}\left(\frac{d-a}{3}\right)\chi(d-a)E_{c_1,m}\left(\frac{d-a}{d}\right)+\frac{1}{1+c_2^{-1}}\mathbf{e}\left(\frac{a}{3}\right)\chi(a)E_{c_2,m}\left(\frac{a}{d}\right)\right) \\
&= \sum_{a=1}^{d-1}\left(\frac{1}{1-\omega^d}\omega^{d-a}(-1)^{\delta(\chi)}\chi(a)(-1)^m E_{c_2,m}\left(\frac{a}{d}\right)+\frac{1}{1-\omega^{2d}}\omega^{2a}\chi(a)E_{c_2,m}\left(\frac{a}{d}\right)\right) \\
&= \frac{[(-1)^{\delta(\chi)+m+1}+1]}{1-\omega^{2d}}\sum_{a=1}^{d-1}\omega^{2a}\chi(a)E_{c_2,m}\left(\frac{a}{d}\right)
\end{aligned}$$

Then, we have that

Theorem 3 *Suppose d is prime to 3, $\omega = \mathbf{e}(1/3)$, and $m \equiv \delta(\chi)-1$ mod 2, then*

$$\begin{aligned}
L(-m,\chi) &= \frac{2d^m}{(\chi(3)3^{m+1}-1)(1-\omega^{2d})}\sum_{a=1}^{d-1}\omega^{2a}\chi(a)E_{-\omega^d,m}\left(\frac{a}{d}\right) \\
&= \frac{2d^m}{(\chi(3)3^{m+1}-1)(1-\omega^2)}\sum_{a=1}^{d-1}\omega^{2a}\chi(a)E_{-\omega,m}\left(\frac{a}{d}\right)
\end{aligned} \tag{3.3}$$

which is another form of [4] Th6.6(i).

References

1. G. Shimura, The critical values of generalizations of the Hurwitz zeta function. Doc. Math. **15**, 489–506 (2010)
2. G. Shimura, The special values of the zeta functions associated with cusp forms. Commun. Pure Appl. Math. **29**, 783–804 (1976) (=Collected Parpers II, 740–761)
3. G. Shimura, The special values of the zeta functions associated with Hilbert modular forms. Duke Math. J. **45**, 637–679 (1978)
4. G. Shimura, *Elementary Dirichlet Series and Modular Forms*, Springer Monographs in Mathematics (Springer, Berlin, 2007)
5. G. Shimura, The critical values of certain Dirichlet series. Doc. Math. **13**, 775–794 (2008)
6. M.J. Dancs, T.X. He, An Euler-type formula for $\zeta(2k+1)$. J. Number Theory **118**, 192–199 (2006)
7. Y. Hamahata, Poly-Euler polynomials and Arakawa-Kaneko type zeta functions. Funct. Approx. Comment. Math. **51**, 7–22 (2014)
8. T. Kim, Euler numbers and polynomials associated with zeta functions. Abstr. Appl. Anal. **581582** (2008)
9. T. Kim, On the generalized degenerate Euler numbers and polynomials. Proc. Jangjeon Math. Soc. **18**, 537–546 (2015)
10. T. Kim, Degenerate generalized q-Euler polynomials. Glob. J. Pure Appl. Math. **11**, 2627–2633 (2015)
11. Q.M. Luo, Some formulas for Apostol-Euler polynomials associated with Hurwitz Zeta function at rational arguments. Appl. Anal. Discret. Math. **3**, 336–346 (2009)
12. M. Lerch, Note sur la fonction $\zeta(\omega, x, s) = \sum_{k=0}^{\infty} e^{2k\pi i x}/(\omega+k)^s$. Acta Math. **11**, 19–24 (1887)
13. A. Hurwitz, Einige eigenschaften der dirichlet'schen functionen $F(s) = \sum(\frac{D}{n} \cdot \frac{1}{n^s})$, die bei der bestimmung der classenanzahlen binärer quadratischer formen auftereten. Zeitschrift für Math. und Phys. **27**, 86–101, (1882) (= Mathematische Werke I, 72–88)

On Orders of Tame Kernels in Quaternion Extension of Number Fields

Haiyan Zhou

Abstract Let E/F be a Galois extension of number fields with Galois quaternion group Q_8, and let K/F be the maximal subextension in E/F. In this paper, we prove that for every odd prime number p and every positive integer m, $2|p^m - \text{rank}\ (K_2(E/K))$, i.e., $2|p^m-\text{rank}\ (K_2\mathcal{O}_E) - p^m-\text{rank}\ (K_2(\mathcal{O}_K))$, and if E is a totally real number field, then $k_2(K)|k_2(E)$, where $k_2(\cdot)$ is the order of the tame kernel of the field $\cdot$.

Keywords Tame kernels · Quaternion group · Transfer map

2010 Mathematics Subject Classification 11R70 · 11R21 · 12F10

1 Introduction

Let F be an algebraic number field, $\mathcal{O}_F$ the ring of integers in F. The Milnor K-group $K_2\mathcal{O}_F$ is known to be the same as the tame kernel of F. The tame kernels of number fields have been investigated by many authors (see, e.g., [5–7, 7–10, 12–23]). Let E/F be a Galois extension with the Galois group G. Applying the Brauer-Kuroda relations between the Dedekind zeta functions of a number field E and of some of its subfields, A. Bartel and B. de Smit [1] obtained some equalities between orders of higher K-groups of E and that of some subfields of E/F for some groups G, for example, elementary abelian p-groups, semidirect products $C_p \rtimes C_n$ with p an odd prime and with C_n acting faithfully on C_p and Heisenberg groups. However, when the Galois group G is a quaternion group Q_8, there do not exist Brauer-Kuroda relations between zeta function of the field E and zeta functions of its proper subfields, one

Supported by NSFC 11471162 and the Natural Science Foundation of the Jiangsu Higher Education Institutions of China (13KJB110016).

H. Zhou (✉)
School of Mathematical Sciences, Nanjing Normal University, Nanjing 210023, People's Republic of China
e-mail: haiyanxiaodong@gmail.com

Z. Liang and C. Aribam (eds.), *The Computational and Theoretical Aspects of Elliptic Curves*, Mathematical Lectures from Peking University,
https://doi.org/10.1007/978-981-13-6664-2_8

may conjecture that the order of the tame kernel of E is not uniformly bounded by orders of tame kernels of its proper subfields. If $E/\mathbb{Q}$ is a Galois extensions with the quaternion Galois group, then we call that E is a quaternion number field. A quaternion number field E with maximal biquadratic subfield K is called pure if discriminants $d(E)$ and $d(K)$ have the same prime divisors. Recently, J. Browkin ([2]) gave some numerical evidence for this conjecture. Assuming the Birch-Tate conjecture for real pure quaternion number fields E, he computed $k_2(E)$ and $k_2(K)$, where $k_2(\cdot)$ is the order of the tame kernel of the field $\cdot$, the field K is the unique maximal subfield of E. From his computation, he observed that $k_2(K)|k_2(E)$ and $k_2(E)/k_2(K)$ is a square. The divisibility of odd parts follows from $(E : K) = 2$. For an odd prime number $p \not\equiv 1 \pmod 8$, Jerzy Browkin ([2]) proved that if $p \nmid k_2(K)$, then the p-primary parts of $k_2(E)$ is a square. For $p \equiv 3 \pmod 4$, the author and Wenzhu Xie proved in [23] that if there is at most one quadratic subfield such that the p-Sylow subgroup of the tame kernel is nontrivial, then the p-primary parts of $k_2(E)/k_2(K)$ is the square of an integer.

Let E/F be a Galois extension of number fields with Galois group Q_8, and let K/F be the maximal subextension in E/F. In the present paper, we prove that for every odd prime number p and every positive integer m, $2|p^m$−rank $(K_2(E/K))$, where $K_2(E/K)$ is the kernel of the map $\mathrm{tr}_{E/K} : K_2\mathcal{O}_E \to K_2\mathcal{O}_K$. Therefore, the p-primary parts of $k_2(E)/k_2(K)$ is the square of an integer. If E is a totally real number field, then $k_2(K)|k_2(E)$. In particular, if E is a totally real quaternion number field, we obtain that $k_2(K)|k_2(E)$ and the odd parts of $k_2(E)/k_2(K)$ is a square.

After completing the final draft of this paper, I was informed by J. Browkin that he had extended his results from [2]. He proved (not assuming any conjecture) that if K is the biquadratic subfield of a totally real quaternion field E, then the odd part of $k_2(E)/k_2(K)$ is a full square by using properties of Artin's L-functions.

2 Main Results

The transfer for K_2 is useful in this section. For the convenience of the reader, we recall its basic properties (see [8]) and some well known facts which we need. Let G be a finite cyclic group and g a generator of G. If A is a G-module, then we write $I_G A = \{ga - a | a \in A\}$, $A_G = A/I_G A$.

Lemma 1 ([8]) *(i) A transfer* $\mathrm{tr}_{E/F}$ *is a group homomorphism* $\mathrm{tr}_{E/F} : K_2(E) \to K_2(F)$.

(ii) The projection formula: for $x \in F$ *and* $y \in E$ *we have*

$$\mathrm{tr}_{E/F}\{x, y\} = \{x, N_{E/F}y\}$$

where $N_{E/F}$ *is just the field norm of the extension* E/F.

(iii) The composite

$$K_2(F) \xrightarrow{j} K_2(E) \xrightarrow{\mathrm{tr}_{E/F}} K_2(F),$$

where j is induces by the inclusion $F \subset E$, is the same as multiplication by the degree of E/F.

(iv) For E/F an extension of number fields and $\mathfrak{p}$ a finite prime of F the following square commutes

$$\begin{array}{ccc} K_2(E) & \xrightarrow{(\tau_{\mathfrak{P}})} & \oplus_{\mathfrak{P}|\mathfrak{p}} k_{\mathfrak{P}}^* \\ \mathrm{tr}_{E/F} \downarrow & & \downarrow (N_{k_{\mathfrak{P}}/k_{\mathfrak{p}}}) \\ K_2(F) & \xrightarrow{\tau_{\mathfrak{p}}} & k_{\mathfrak{p}}^* \end{array}$$

where $\mathfrak{P}$ is a prime ideal of E, $k_{\mathfrak{P}}$ the residue field of E, $k_{\mathfrak{p}}$ the residue field of F, $\tau_{\mathfrak{p}}$ the homomorphism induced by the tame symbols on F and $\tau_{\mathfrak{P}}$ the homomorphism induced by the tame symbols on E.

(v) For E/F a Galois extension with group G and a positive integer n the homomorphism $j : K_2(F) \to K_2(E)$ induces a homomorphism $j : {}_nK_2(F) \to {}_nK_2(E)$ which is an injection when n and $|G|$ are relatively prime. At this time the $tr_{E/F}$ induces a surjection $\mathrm{tr} : {}_nK_2(E) \to {}_nK_2(F)$.

(vi) Suppose that E/F is a Galois extension with Galois group G, we have $j\mathrm{tr}_{E/F} = N_{E/F}$, where $N_{E/F}$ is the group norm, $N_{E/F}(x) = \prod_{\sigma \in G} \sigma(x)$.

(vii) If $j : K_2(F) \to K_2(E)$ and $\mathrm{tr}_{E/F} : K_2(E) \to K_2(F)$ are restricted to the groups $K_2\mathcal{O}_E$, $K_2\mathcal{O}_F$, then the analogues of (v) and (vi) hold for these groups as well.

Let T be a set of primes of E containing the infinite primes of E. Then for rings of T-integers in E, we have short exact sequences

$$0 \to K_2(\mathcal{O}_{E,T}) \to K_2(E) \to \bigoplus_{\mathfrak{p} \notin T} k_{\mathfrak{p}}^* \to 0 \tag{1}$$

and

$$0 \to K_2(\mathcal{O}_E) \to K_2(\mathcal{O}_{E,T}) \to \bigoplus_{\mathfrak{p} \in T, \mathfrak{p} \text{ finite}} k_{\mathfrak{p}}^* \to 0. \tag{2}$$

For a number field E, $K_2^+(E)$ is defined as the kernel of

$$(\lambda_{\mathfrak{p}}) : K_2(E) \to \bigoplus_{\mathfrak{p} \text{ real infinite}} \mu_2,$$

where $\lambda_{\mathfrak{p}}$ is induced by the Hilbert symbol $(,/\mathfrak{p})_2$ on the real number field $\mathbb{R}$ and $\mu_2 = \{\pm 1\}$. It is well known that for each real embedding of E there exists an $\alpha \in E^*$

negative under this embedding and positive under the others. Then we have a split short exact sequence

$$0 \to K_2^+(E) \to K_2(E) \to \bigoplus_{\mathfrak{p} \text{ real infinite}} \mu_2 \to 0. \tag{3}$$

For rings of T-integers in E, the group $K_2^+(\mathcal{O}_{E,T})$ is defined as $K_2^+(\mathcal{O}_{E,T}) = K_2(\mathcal{O}_{E,T}) \cap K_2^+(E)$. Moreover, we have a short exact sequence

$$0 \to K_2^+(\mathcal{O}_E) \to K_2\mathcal{O}_E \to \bigoplus_{\mathfrak{p} \text{ real infinite}} \mu_2 \to 0. \tag{4}$$

Lemma 2 ([8, Lemma 2.3]) *The homomorphism*

$$K_2(E) \to (\bigoplus_{\mathfrak{p}\, finite} k_{\mathfrak{p}}^*) \bigoplus (\bigoplus_{\mathfrak{p}\, real\ infinite} \mu_2)$$

induced by the $\tau_{\mathfrak{p}} : K_2(E) \to k_{\mathfrak{p}}^*$ and $\lambda_{\mathfrak{p}} : K_2(E) \to \mu_2$ is surjective.

By Lemmas 2 and (2), we have a short exact sequence

$$0 \to K_2^+(\mathcal{O}_E) \to K_2^+(\mathcal{O}_{E,T}) \to \bigoplus_{\mathfrak{p} \in T, \mathfrak{p} \text{ finite}} k_{\mathfrak{p}}^* \to 0. \tag{5}$$

Lemma 3 ([8, Proposition 6.2]) *Let E/F be a Galois extension of number fields with a solvable Galois group G. Let S be a set of primes of F containing the infinite primes and also the finite primes $\mathfrak{p}$ for which $N(\mathfrak{p}) - 1$ and the ramification index $e_{\mathfrak{p}}$ of $\mathfrak{p}$ in E/F are not relatively prime, and let T be the primes of E which are above the primes in S. Then $tr_{E/F} : K_2(E) \to K_2(F)$ induces an isomorphism*

$$\text{tr} : K_2^+(\mathcal{O}_{E,T})_G \to K_2^+(\mathcal{O}_{F,S}).$$

Let E/F be an extension of number fields. Denote by $K_2(E/F)$ and $K_2^+(E/F)$ the kernel of the map $\text{tr}_{E/F} : K_2\mathcal{O}_E \to K_2\mathcal{O}_F$ and $\text{tr}_{E/F} : K_2^+(\mathcal{O}_E) \to K_2^+(\mathcal{O}_F)$, respectively.

Theorem 1 *Let E/K be a quadratic extension of number fields with a Galois group G. Then $tr_{E/K} : K_2(E) \to K_2(K)$ induces a surjective* tr : $K_2^+(\mathcal{O}_E) \to K_2^+(\mathcal{O}_K)$, *and $K_2^+(E/K) = I_G K_2^+(\mathcal{O}_{E,T})$, where S is a set of primes of K containing the infinite primes and also the finite primes $\mathfrak{p} \nmid 2$ for which $\mathfrak{p}$ in E/K are totally ramified, and T is the primes of E which are above the primes in S.*

Proof By Lemma 1(iv) and (5), we consider the following commutative diagram:

$$\begin{array}{ccccccccc} 0 & \longrightarrow & K_2^+(\mathcal{O}_E) & \longrightarrow & K_2^+(\mathcal{O}_{E,T}) & \longrightarrow & \bigoplus_{\mathfrak{P}\in T, \mathfrak{P} \text{ finite}} k_{\mathfrak{P}}^* & \longrightarrow & 0 \\ & & \downarrow \text{tr} & & \downarrow \text{tr}_{E/K} & & \downarrow (N_{k_{\mathfrak{P}}^*/k_{\mathfrak{p}}^*}) & & \\ 0 & \longrightarrow & K_2^+(\mathcal{O}_K) & \longrightarrow & K_2^+(\mathcal{O}_{K,S}) & \longrightarrow & \bigoplus_{\mathfrak{p}\in S, \mathfrak{p} \text{ finite}} k_{\mathfrak{p}}^* & \longrightarrow & 0 \end{array},$$

where tr is the restriction of $\text{tr}_{E/K}$, $k_{\mathfrak{P}}$ the residue field of E and $k_{\mathfrak{p}}$ the residue field of K. It is easy to see that $N_{k_{\mathfrak{P}}^*/k_{\mathfrak{p}}^*}$ is the identity mapping since $(E : K) = 2$. By the snake lemma, we have tr is surjective and ker(tr)= ker($\text{tr}_{E/K}$). Therefore, this results follows from Lemma 3.

Corollary 1 *Let E/K be a quadratic extension of number fields with a Galois group G. If E is a totally real number field, then $k_2(K)|k_2(E)$.*

Proof Since E is a totally real number field, we know that K is also totally real, and $r_1(E) = 2r_1(K)$. By Theorem 1, we have $|K_2^+(\mathcal{O}_K)|||K_2^+(\mathcal{O}_E)|$. Moreover, we have $k_2(E) = 2^{r_1(E)}|K_2^+(\mathcal{O}_E)|$ and $k_2(K) = 2^{r_1(K)}|K_2^+(\mathcal{O}_K)|$ by (4). Therefore, we have $k_2(K)|k_2(E)$.

Corollary 2 *Let E/F be a Galois extension of number fields with a quaternion Galois group Q_8, and let K/F be the maximal subextension in E/F. If E is a totally real number field, then $k_2(K)|k_2(E)$.*

Theorem 2 *Let E/F be a Galois extension of number fields with Galois group Q_8, and let K/F be the maximal subextension in E/F. Then for every odd prime number p and every positive integer m, we have $2|p^m$−rank $(K_2(E/K))$.*

Proof It is clear that $K_2^+(\mathcal{O}_E)(p) = K_2\mathcal{O}_E(p)$ by (4), so $K_2^+(E/K)(p) \cong K_2(E/K)(p)$. Since $\sigma(1-\sigma^2) = (1-\sigma^2)\sigma$ and $\tau(1-\sigma^2) = \tau(1-\tau^2) = (1-\sigma^2)\tau$, we know that $K_2(E/K)(p)$ is a Q_8−module by Theorem 1. Put $V := K_2(E/K)^{p^{m-1}}/K_2(E/K)^{p^m}$. Then the group Q_8 acts on the elementary abelian p-group V since $K_2(E/K)(p)$ is a Q_8-module. Let Φ denotes the $\mathbb{F}_p$-representation $\Phi : Q_8 \to \text{Aut}V$ induced by the action of Q_8 on V. Firstly, we will prove that if $V \neq 1$, then Φ is faithful. It is clear that kerΦ is a normal subgroup of Q_8 which acts trivially on V. By Galois Theory, there is a normal subextension L/F of E/F corresponding to kerΦ. Then $L \subseteq K$, so we have $N_{E/K}(v) = v^2$ for every $v \in V$. By assumption and Lemma 1(vi), we conclude that $v^2 = N_{E/K}(v) = j\text{tr}_{E/K}(v) = 1$.

If $V \neq 1$, we can find a element $v \neq 1 \in V$. Then v is killed by 2 and p. It follows that $v = 1$. This is a contradiction. Therefore, Φ is faithful.

Secondly, since p is an odd prime number, we know that every $\mathbb{F}_p[Q_8]$-module is completely reducible by Maschke's theorem (See [6, Theorem 1.9]). Then the $\mathbb{F}_p$-character χ of Q_8 afforded by Φ is the sum of characters of Q_8 afforded by irreducible representations. It is obvious that p^m−rank $K_2(E/K) = \chi(1)$. Since the degree of character χ of the only irreducible faithful $\mathbb{F}_p$-representations of Q_8 is divisible by 2 (see [11, P_{420}]), we have $2|\chi(1)$, i.e., $2|p^m$−rank $K_2(E/K)$.

Corollary 3 *Let E/F be a Galois extension of number fields with a quaternion Galois group Q_8. Let K/F be the maximal subextension in E/F. Then the odd parts of $k_2(E)/k_2(K)$ is a square.*

Unfortunately, we can not get any results about 2^m-rank of $K_2(E/K)$. However, we can obtain the following results for the 2-primary part of $K_2^+(E/K)$.

Proposition 1 *Let E/F be a Galois extension of number fields with a quaternion Galois group Q_8. Let K/F be the maximal subextension in E/F. If $K_2^+(E/K)(2) \cong (2^\alpha, 2^\beta, \cdots)$, where $\alpha \geq \beta \geq \cdots$, then $\alpha - \beta \in \{0, 1\}$.*

Proof Let $a \in K_2^+(E/K)$ be an element of order 2^α. Then a^{2^β} generates a subgroup A of $K_2^+(E/K)$ of order $2^{\alpha-\beta}$. Since $a^\sigma = a^i b$ for some odd integer i and an element d of order dividing 2^β, we obtain that A is a Q_8-module. The rest is to prove $|A||2$. Assume that A has an element c of order 4. Then the order of c^σ is also 4, hence we must have $c^\sigma = c$ or $c^\sigma = c^3$. In both cases we have $c^{\sigma^2} = c$, then $1 = (j\mathrm{tr})(c) = N_{E/K}(c) = c^{1+\sigma^2} = c^2$, contradicting our assumption. Therefore, A is elementary abelian. It implies $|A||2$ since A is cyclic. This proves our claim.

Acknowledgements The author is grateful to Professor Jerzy Browkin for giving her his paper [2] and correspondence.

References

1. A. Bartel, B. de Smit, Index formulae for integral Galois modules. J. Lond. Math. Soc. **88**, 845–859 (2013)
2. J. Browkin, K-groups of quaternion number fields, in *Conference on Algebraic K-Theory and Arithmetic in Honour of Jürgen Hurrelbrink* (Banach Center, Bedlewo, Poland, 2012)
3. J. Browkin, Tame kernels of cubic cyclic fields. Math. Comput. **74**, 967–999 (2005)
4. J. Coates, On K_2 and some classical conjectures in algebraic number theory. Ann. Math. **95**, 99–116 (1972)
5. H. Garland, A finiteness theorem for K_2 of a number field. Ann. Math. **94**, 534–548 (1971)
6. I.M. Isaacs, *Character Theory of Finite Groups* (Academic, Cambridge, 1976); 2nd edn. (Dover, New York, 1995)
7. B. Kahn, Descente galoisienne et K_2 des corps de nombres. K-theory **7**, 55–100 (1993)
8. F. Keune, On the structure of the K_2 of the ring of integers in a number field. K-Theory **2**, 625–645 (1989)
9. M. Kolster, Odd torsion in the tame kernel of totally real number fields, in *Algebra K-Theory: Connections with Geometry and Topology*. NATO Adv. Sci. Inst. Ser. C Math. Phys. Sci., vol. 279 (Kluwer Academic Publishers, Dordrecht, 1989), pp. 177–188
10. Y.Y. Li, H.R. Qin, On the 3-rank of tame kernels of certain pure cubic number fields. Sci. China Math. **53**(9), 2381–2394 (2010)
11. G. James, M. Liebeck, *Representations and Characters of Groups*, 2nd edn. (Cambridge University Press, Cambridge, 2001)
12. H.R. Qin, The 2-Sylow subgroups of the tame kernel of imaginary quadratic fields. Acta Arith. **69**, 153–169 (1995)
13. H.R. Qin, The 4-rank of $K_2\mathcal{O}_F$ for real quadratic fields. Acta Arith. **72**, 323–333 (1995)
14. H.R. Qin, The structure of the tame kernels of quadratic number fields (I). Acta Arith. **113**, 203–240 (2004)

15. H.R. Qin, The 2-Sylow subgroup of $K_2\mathcal{O}_F$ for number fields F. J. Algebra **28**, 494–519 (2005)
16. H.R. Qin, Tame kernels and Tate kernels of quadratic number fields. J. Reine Angew. Math. **530**, 105–144 (2001)
17. H.R. Qin, H.Y. Zhou, The 3-Sylow subgroup of the tame kernel of real number fields. J. Pure Appl. Algebra **209**, 245–253 (2007)
18. D. Quillen, *Finite Generation of the Groups* K_i *of Rings of Algebraic Integers*. Lecture Notes in Mathematics, vol. 341 (Springer, Berlin, 1973), pp. 179–198
19. C. Soulé, Groupes de Chow et K-théorie de variétés sur un corps fini. Math. Ann. **268**, 317–345 (1984)
20. X. Wu, Tame kernels of quintic cyclic fields. Acta Arith. **134**, 183–199 (2008)
21. H. Zhou, The tame kernel of multiquadratic number fields. Commun. Algebra **37**, 630–638 (2009)
22. H. Zhou, Tame kernels of cubic cyclic fields. Acta Arith. **124**, 293–313 (2006)
23. H. Zhou, W. Xie, Tame Kernels of quaternion number fields. Commun. Algebra, **42**, 2496–2501, 630–638 (2014)

The manufacturer's authorised representative in the EU is Springer Nature Customer Service Centre GmbH, Europaplatz 3, 69115 Heidelberg, Germany. If you have any concerns regarding our products, please contact ProductSafety@springernature.com

Printed and bound by CPI Group (UK) Ltd, Croydon, CR0 4YY
25/09/2026
02222898-0017